FARM MACHINERY MANAGEMENT

FARM MACHINERY MANAGEMENT

by
Kartik Basu | Ashish Anne

Wordpen Academics
Publishing Science That Matter

ISBN: 9789360843359

Published by : WORDPEN ACADEMICS
4736/23, Ansari Road, Darya Ganj,
New Delhi – 110 002
Phone: +91-011-35004336
E-mail: info@wordpenacademics.com

Printed at : Replika Press Pvt. Ltd.

Exclusive Distributor: ***Bio-Green Books, New Delhi***

PREFACE

In modern agriculture, the efficient management of farm machinery stands as a cornerstone of productivity, profitability, and sustainability. From plowing fields to harvesting crops, farm machinery plays a crucial role in every stage of the agricultural production process, significantly impacting the efficiency and effectiveness of farming operations. As farms become increasingly mechanized and technologically advanced, the proper selection, operation, and maintenance of farm machinery become essential skills for farmers and agricultural professionals alike.

This book, titled "Farm Machinery Management," is designed as a comprehensive guide to understanding the principles, practices, and strategies for effective farm machinery management. Drawing upon the expertise of agricultural engineers, farm managers, and machinery specialists, this volume offers insights, techniques, and best practices for optimizing the use of farm machinery in agricultural operations.

From the selection and procurement of farm machinery to the operation, maintenance, and replacement decisions, "Farm Machinery Management" covers a wide range of topics essential for maximizing the efficiency and productivity of farming enterprises. Through practical examples, case studies, and hands-on exercises, readers will gain a deep understanding of the factors influencing machinery management decisions and the strategies for overcoming common challenges in farm machinery operations.

Moreover, this book explores emerging trends and technologies in farm machinery management, including precision agriculture, autonomous vehicles, and digital tools for equipment monitoring and optimization. By embracing innovation and technological advancements, farmers can enhance decision-making processes, reduce operational costs, and minimize environmental impact while maximizing yields and profitability.

As we navigate the complexities of farm machinery management, it is our hope that "Farm Machinery Management" will serve as a valuable resource for farmers, agricultural professionals, educators, and students seeking to enhance their knowledge and skills in this critical area. Whether you are a seasoned farmer looking to improve machinery efficiency on your operation or a student exploring career opportunities in agriculture, this book invites you to delve into the world of farm machinery management and discover the tools and techniques for success in modern farming.

Kartik Basu | Ashish Anne

CONTENTS

1

INTRODUCTION

Agricultural engineering is the engineering discipline that applies engineering science and technology to agricultural production and processing. Agricultural engineering combines the disciplines of animal biology, plant biology, and mechanical, civil, electrical and chemical engineering principles with a knowledge of agricultural principles.

Although agricultural engineering is a fairly new subject, its effect is being felt in India increasingly. In recent years it has contributed to agricultural production in the country both directly and indirectly. The number of modern machines used in agriculture is a fairly good index of the progress made during the four 5 year plans. And yet all this progress is insignificant as compared to that in USSR, USA, Japan, UK Canada, and Australia. This development has made two things clear: 1. That the Indian farmers are quick in adopting new methods and techniques and 2. That the Indian industries can manage as good machines as those abroad.

Agricultural engineering as defined by international institutions comprises of four main branches, (i)farm implements (ii)rural structures (iii)soil conservation, drainage, and irrigation and (iv)rural electricity.

Because of the peculiar local conditions the lack of electricity on a large scale, and the immediate utility of implements, agricultural engineering in India is more of agricultural implements and machinery. But in fact agricultural engineering is a very vast subject and may be defined as the application of the knowledge, techniques and disciplines of various fields of engineering to the solution of problems arising in the field of agriculture and rural living with the object of reducing labour, improving productivity and raising the standard of living of the farmer and increasing the overall earnings per worker. The era of more intensive planning brought re-organisation on a wider scale. With the establishment of agricultural universities, several agricultural engineering colleges were established offering graduate and post graduate courses. The most important step has been the establishment of agro-industries corporations with the financial assistance of the Central Ministry of Agriculture. The main objectives of establishing these corporations have been to take over the supply and service functions of the departments of agriculture and to expand them. These agri-industries corporations have each a paid-up capital of 2-5 crores of rupees. They have taken over the existing government workshops also.

The other most important development has been in the field of industry which has so far half a dozen factories for manufacturing tractors, 3 or 4 for power tillers, and several for oil engines and electrically driven pumpimg sets. Some of these are being assisted by the Industrial Finance Corporation of India. These helped to quicken the pace of mechanisation of Indian agriculture.

It is not merely the knowledge of agricultural engineering which is needed for increasing production, but also the knowledge of their availability, prices and upkeep. Information can be obtained from:

1. At the village level: the village level worker or gram sevak.
2. At the block level: the Agricultural Extension Officer or the District Planning Officer.
3. At the district level: The District Agricultural Officer or the District Planning Officer.
4. At the state level: the State Agricultural Engineering Section and the Managing Director of the State Agro-Industries Corporation.
5. At the Central level: the Director of Agricultural implements and Joint Commissioner (machinery), Ministry of Agriculture and Irrigation, Krishi Bhavan, New Delhi.

The private industry as well as the government, through their agro-industries corporations are in the field now for the supply of agricultural machinery to farmers. To prevent the charging of exorbitant prices for tractors, the govt has fixed the price of tractors with the assistance of the Bureau of Cost Accounting in the Ministry of Finance. This organisation has helped the farmers very much.

Agricultural aviation is now gaining ground in India. At present the aeroplanes and helicopters are being used mainly for spraying pesticides but in future they may be used for sowing and spraying liquid fertilizers on the crops. Agricultural electronics has also begun and remote-control tractors are being designed and tested in some countries. With the successful explosion of an atomic device India may be able to generate more electricity in the future. The availability of plentiful irrigation water as a result of the Ganga-Kauveri project and the availability of more electricity from atomic energy will make the progress of farm mechanisation in India speedier. Thus the future seems to be bright.

Agricultural engineering is an engineering discipline that applies engineering science and technology to agricultural production and processing. Agricultural engineering combines the disciplines of animal biology, plant biology, mechanical, civil and chemical engineering principles with knowledge of agricultural principles.

When the knowledge of engineering and biological science is applied to agriculture, it is called agricultural engineering. Agricultural engineers are responsible for designing agricultural tools and equipment as well as machinery and plants. These engineers can either specialize in designing power and machine systems (for environmental or agricultural purposes) or be involved in food and bioprocess. Soil and water conservation as well as agricultural processing are some of their concerns. Agricultural engineers may work in R&D operation or sales or management.

HISTORY

The first curriculum in Agricultural Engineering was established at Iowa State University by J. B. Davidson in 1905. The American Society of Agricultural Engineers, now known as the American Society of Agricultural and Biological Engineers, was founded in 1907.

Agricultural Engineers

Agricultural Engineers may perform tasks as planning, supervising and managing the building of dairy effluent schemes, irrigation, drainage, flood and water control systems, perform environmental impact assessments, agricultural product processing and interpret research results and implement relevant practices. A large percentage of agricultural engineers work in academia or for government agencies such as the United States Department of Agriculture or state agricultural extension services. Some are consultants, employed by private engineering firms, while others work in industry, for manufacturers of agricultural machinery, equipment, processing technology, and structures for housing livestock and storing crops. Agricultural engineers work in production, sales,management, research and development, or applied science.

In the United Kingdom the term Agricultural Engineer is often also used to describe a person that repairs or modifies agricultural equipment.

Subfields

Some of the specialties of agricultural engineers include:

- the design of agricultural machinery, equipment, and agricultural structures

- “crop production, including seeding, tillage, irrigation and the conservation of soil and water
- “animal production, including the care and processing of poultry and fish and dairy management
- “the processing of food and other agricultural and bio-renewable products,and food engineering.
- “Bioresource engineering, which uses machines on the molecular level to help the environment.

AGRICULTURAL IMPLEMENTS AND MACHINERY

The beginning of mechanisation of Indian agriculture was made by the use of improved hand tools and improved bullock-drawn implements. Particular stress was laid on these items during the first 3 Five year plans and slowly more sophisticated implements were introduced. These included power-driven threshers, tractors and the combine harvestors. Quite a variety of implements is required for carrying agricultural operations , starting from thr opening of the land to harvesting, threshing, winnowing and storage. These operations include ploughing, harrowing, levelling, sowing, intercultivation, application of fertilizers and manures, harvesting, threshing and winnowing.

A variety of desi implements are in use in the country for the last so many centuries. before efforts were made to introduce foreign implements a survey implements used in India was taken. Therefore under the Indian Council of Agricultural Research a country wide survey was carried out and its report was published. This has brought many facts to light. The number of implements used by the Indian farmer is very large. Some of these implements are crude to look at but are very useful and the consensus of the agricultural engineers of india has been that it would be easier at least as a first step to improve upon the indigenous implements in order to increase their efficincy and thus give to the Indian farmers something which is within their means.

To achieve this objective effectively, the Indian Council of Agricultural research has now established one research-cum testing and trining centre in each state. Some of them have now been converted into regional centres or have been taken over by the state governments. The object is to improve the indigenous agricultural implements, to design new implements and to test them in the field upto the prototype level. After the implements have proved to be useful, the prototypes can be given to manufacture in public or private sectors and the implements can be sold through them to the farmers.

Principles of research. In India, the holdings are generally small, the average farmer is poor, and the draft cattle are also small and ill fed. Taking these into account certain principles have been laid down which are:

1. That the implements and machinery for the Indian farmers should be simple in construction, so that they could be operated by the illiterate farmers, and should either be manufactured could be repaired by the village artisans or mechanics.
2. The prices of the implements should be within reach of the ordinary farmer. In case of power-driven equipment the difficulty of high cost can be overcome by giving the machinery on hire or hire-purchase system or on custom basis or on co-operative basis.
3. The implements should be light so that they can be easily transported and are suitable for the draft animals.
4. That the implements should be scientifically tested on a fairly uniform bais throughout India.
5. In designing the implements the local available materials are to be used. In the case of soil working and plant working parts durable materials such as carbon steel should be used.

To fulfil these requirements, the Indian Council of Agricultural Research has also designed procedures for testing and designing implements and proformas in which the test data are to be listed. Quite a large number of farmers are using the power machines like tractors, oil engines and pumps. Thus the trend towards mechanised farming is bound to increase.

Recently, new developments have taken place to bring about mechanization of agriculture. Owing to the shortage of diesel oil and its increased price, the mechanical operation will have to be restricted to deep ploughing, land levelling, clearing land and other operations which cannot be carried out by bullock power. For short distances bullock carts are more effective than trucks;hence the necessity for improving them.

Agricultural engineering, like all other branches of engineering uses scientific principles and technology to solve problems relating to sustainable agricultural production. Agricultural Engineers may perform tasks such as planning, supervising and managing the building of dairy effluent schemes, irrigation, drainage, flood and water control systems, perform environmental impact assessments and interpret research results and implement relevant Practices.

Agricultural engineers also specialize in other aspects like power system and machinery design; structures and environmental science; and food and bioprocess engineering. They discover the ways to conserve soil and water and to improve the processing of agricultural products. Agricultural Engineering Division Board (AGDB) and Agricultural Engineering Division of Karnataka State Centre is organising lEI Colloquium of Agricultural Engineers as part of 26th Indian Engineering Congress - 2011

This Division was established in 1978

Agricultural engineering integrates the principles of try further poses a problem since they often interact with Biological science and Technological Engineering into wide spectrum of disciplines and with society.

Agricultural production as well as processing

The history of Indian Space activities can be traced to the establishment of TERLS in 1963 for studies of the upper atmosphere and ionosphere over the magnetic equator passing through Thumba near Thiruvananthapuram using small sounding rockets.

Need of the hour warrants stepping up agricultural and allied Specialties include, power systems, machinery production thus contributing an increased & significant design, structures and environment, bioprocess share in the GDP which can be made possible by increasing engineering, management of dairy effluent system. mechanization and precision agriculture in order to ensure Challenges that they meet are storage solution for judicious use of resources.

Grains, live stock, enhancing use of non-conventional energy sources in agriculture and allied sector, soil & Participation water conservation, rainwater harvest and recycling, agriculture drainage & ground water recharge and Professionals, academicians, scientists, engineers and appropriate implement & machinery system. The students who are all involved in the growth as well as interaction with growers is a minimum requirement progress of our Nation, have an opportunity to avail this and climate changes causes major challenges. 70 to platform for exchange of ideas. Please visit the website for 75% ofthe rainwater is utilized in this sector. Alter- registration and other details. Native uses for agricultural by products is to be Papers for Colloquium are invited on the above mentioned reviewed with its impact on ecology and environ- thrust areas.

Agricultural Engineering Activities

Mechanisation plays very important role in the field of Agriculture. The timeliness of operation with an ease and economy are the important parameters for increased production. Intensive cultivation is only possible by using the farm-machineries. The use of improved Agricultural equipments can increase the production by 15-20 percent. More than 50 % of the power in the state comes from animals; mechanization level needs to be increased looking to the crop diversification

& multicropping programmes of the state. Department of Agriculture-Engineering provides custom-hiring facilities to the cultivators on reasonable rates, as the small & marginal farmers can not afford to purchase the costly machines like tractors/ Dozers and the new equipments.

Following activities are being taken up by the department :-

1. Custom-hiring for Land levelling, bunding, dozing and preparation of tanks and other land improvement activities by dozers, cultivation works like seed-bed preparation, sowing, harvesting, threshing etc by departmental tractors, reapers & threshers etc.
2. Shakambhari scheme has been started in the year 2005-06 under which 75% subsidy is being given to small & marginal farmers on Diesel/electric/ Kerosene pump sets (Upto 5HP) and 50% on dugwells.
3. Live-Demonstrations of improved implements/machinery for popularization of the implements amongst the farmers.
4. Popularization of mechanization is an important issue at GOI level and different crop-wise Schemes for popularizing improved Bullock/Tractor drawn implements, Power tiller, tractors and hand tools etc is being taken up. State Govt. is giving an equal additional subsidy on all the items except tractors and combine Harvester.
5. Distribution of Tractors, Power Tillers, Power/Tractor/Bullock drawn implements and Hand Tools is done by C.G. State Seed and Agri. Development Corporation and suppliers registered at state/distt. level in the deptt. of agri.
6. Training is imparted to the farmers on selection, operation and Maintenance of agricultural machinery/ implements under Rashtriya Krishi Vikas Yojna.
7. Low-lift pumps are distributed to small & marginal on 75% subsidy.

For the implementation of these programmes there are three Divisional offices at Raipur, Bilaspur & Jagdalpur along with one Tractor units at Kondagaon,. Four new Asstt. Agril. Engineer offices are at Kanker, Ambikapur, Raigarh and kabirdham

Agricultural engineers help feed the world by combining scientific and engineering knowledge to find unique solutions that make our natural resources more usable while conserving and replenishing them and protecting our environment.

Agricultural engineers apply engineering principles and biological sciences to designing systems that produce food, feed, fiber, and other useful products from renewable bioresources. At the same time they strive to protect the environment by conserving and replenishing our natural resources. Thus, agricultural engineers integrate the biological and environmental sciences with the engineering profession.

By virtue of their education, training, and experience, agricultural engineers are uniquely qualified to design the equipment, systems, and methods that continue the technological advancement in food and agribusiness industries. This includes the production, processing, handling, and distribution of agricultural products throughout the entire food and agribusiness chain.

The Work of Agricultural Engineers

Agricultural, garden machinery and construction equipment is distributed through machinery dealerships who also maintain and supply parts for the machinery throughout its working life.

Employment in a dealership is usually in five main areas: Service; Parts; Sales; Management and Administration, (including accounts). Service engineers are usually trained through an apprenticeship in conjunction with part-time college and manufacturers courses. They are required to work on a range of machines depending on the season.

Some of the work will be in the workshop at the dealership but much time will be spent repairing machines in the field. Some service engineers specialize in the installation of fixed equipment on farms, such as grain storage buildings, air conditioning systems and milking equipment.

The construction equipment industry is more specialized and usually concentrated around specific manufacturers. Parts specialists are normally trained within the company with the help of external and manufacturers courses. They are responsible for the ordering and sale of parts to ensure that these are readily available when required. Normally stock control systems are computerized but the parts-person has an important role to play in customer relations and needs a very good knowledge of how the machines function.

Managers within a dealership usually progress through the sales or service departments and are responsible for the profitable operation of a branch or a department. They are normally selected on a combination of their ability to lead and their technical experience. Administrators and accountants are normally recruited from outside the industry and their skills adapted to the trade.

Engineers in manufacturing

Farmers and growers need the support of a sophisticated engineering industry to produce the machines and equipment they require to ensure that the crops and livestock they grow reach the high quality standards they must attain to obtain the best market prices.

The construction industry similarly requires a specialist knowledge base to reduce expensive machine downtime. The machinery and equipment manufacturers therefore need engineers of the highest caliber having the skill to deal with structures, mechanisms, control systems, hydraulics and electronics as well as having an interest in and appreciation of farming systems or other relevant industry sectors.

Companies making equipment for agriculture and horticulture vary in size from the large international corporations to small companies specializing in products for a limited sector of the market. With one notable exception construction equipment manufacturers are dominated by the larger internationally based companies. Engineers in the land based sector get involved in the following: tractors and tillage machines, harvesting equipment, crop processing, animal welfare (handling and transport), irrigation and drainage schemes, earth moving and other construction equipment, pioneer road and bridge construction, forestry machines, horticultural machines, and fish farming equipment.

RESEARCH DEVELOPMENT & TEACHING

Research, development and teaching in agricultural engineering is carried out in four types of organization. These are specialist research institutes, the universities, colleges of higher education and the equipment manufacturers.

Research institutes employ graduates and technicians from many disciplines on projects of a fundamental nature. Many are concerned with the problems of environmental protection, food safety, pollution and animal welfare. Other projects are concerned with the introduction of new technologies such as robotics and satellite imagery to agriculture.

The research interests of the universities are very broad ranging and mostly sponsored by government, charitable trusts or industry. Universities may also provide opportunities for academic staff to undertake consultancy work and will have a good honors degree as a minimum. Teaching posts in colleges of further education are mainly concerned with courses at technician and craft levels orientated towards providing practical engineering skills.

Equipment manufacturers employ development and test engineers to ensure that they do not fall behind in the development of new models of proven reliability. These engineers work in close liaison with the designers reporting back any weaknesses that are found during testing and assisting in finding solutions to any problems.

Agricultural Engineering applies the science and technology to agricultural production and processing. It combines the principles of animal biology, plant biology and civil, mechanical and chemical engineering principles with the knowledge of agricultural principles. Agriculture is given the status of priority sector in the Indian economy because directly or indirectly more than 70% of the total work force is employed in this sector and it contributes to around 18% to the Gross Domestic Product. Former Prime Minister of India Pandit Jawaharlal Nehru once said that everything can wait but not agriculture as almost all rural population depends on agriculture and allied activities like animal husbandry, dairy, poultry and fisheries. India is traditionally an agricultural country hence the growth and development of Indian economy largely depends upon continuous growth of agriculture.

Agricultural Engineers use their plethora of knowledge to solve the problems of both farmers as well as agricultural industry. They are responsible for manufacturing machinery for irrigation, drainage systems, floods and water control systems, perform environmental impact assessments, agricultural product processing and interpret research results and implement relevant practices.

Agricultural engineers utilize engineering technology and science for agricultural use by managing biological resources efficiently. Apart from creating new techniques in farming and agriculture, they also apply engineering design and analysis to protect animals and natural resources.

As an agricultural engineer, a person accepts the challenge to improve agricultural production by means of better engineering methods, inventions, technology and equipment. This requires a good intellect and a desire to make a difference to future of farming. The agricultural engineers invent new techniques to conserve soil and water and to improve the processing of agricultural products into value added agriculture products.

Educational Requirement: - A student has to pass the 10th standard board examination with 50% successfully to be eligible for a Diploma course. But if the student really dreams for a bright career in Agricultural Engineering then the need is go for a Bachelor's degree -B.E/B.Tech. The eligibility criterion for a B.E./B.Tech is 50% marks in class 12th Board Examination with Physics, Chemistry and Maths as compulsory subjects and preferably Biology. Diploma Holders in Agricultural Engineering too can get direct admission in the second year of B.Tech course of Agricultural Engineering. A bachelor's degree is required for an advanced course , M.E./M.Tech. A person can also aim for a PhD to do research on matters related with Agricultural Engineering. There is also a provision of Distance Education which helps in attaining a bachelor's degree through Associate Examination of the Institute of Engineers (AMIE)

Selection: - The admission to the agricultural engineering course is very competitive. A student should pass IIT-JEE (Joint Entrance Examination) to attain admission in one of the prestigious IITs. This highly competitive test, checks student's ability in Physics, Chemistry and Mathematics and for other reputed institutions like NIT, one has to succeed in AIEEE(All India Engineering/ Pharmacy/Architecture). Also, there is state level and national level examinations as some institutions opt for their own entrance examinations.

Different Branches

The field of Agricultural is very interesting. If a student really wishes for a career in this field, the need would be to apply oneself. There are different disciplines of Agricultural Engineering and a person can select the right area according to the aptitude and no wonder the person can rise and shine in his professional life.

The various disciplines of agricultural engineering are -

1. **Bioenvironmental Engineering** - This discipline deals with the application of engineering principles to design, manufacture and test systems that provide the desired environmental conditions for animals, human housing, crop storage structures, green houses and other biological systems. Bioenvironmental engineers design equipment for heating, ventilating, air conditioning air quality control, and air quality control. They also develop systems to handle and invent treatment for bio-waste material.

2. **Food and Bioprocess Engineering** - This discipline of agricultural engineering deals with the application of engineering principles to the production, processing and packaging of food and bio-products. Food and bioprocess engineers design, develop and construct new processes, machines and plants. They develop and test new products as well as manage environmental factors, waste products and energy conservation.

3. **Soil and Water Resources Engineering** - Soil and Water Resource Engineering is the application of engineering principles and its practices to design and develop systems for natural resources. It also includes protection of environment by utilizing contour and other appropriate cultivation systems. Soil and water engineers design systems to control soil erosion, flooding and also develop irrigation, drainage systems. They also develop policies to handle storm water and control the movement of sediment in water systems.

4. **Off-Road Equipment Engineering** - Off-Road Equipment Engineering is the application of the principles of Engineering to design, manufacture and test equipment for the agricultural, construction, forestry and mining industries. Off-Road Equipment engineers develop precision machine systems that rely on advanced information, sensing technologies and a high level of automation control.

5. **Agricultural Building and Structures** - Agricultural engineering also involves Agricultural Engineers in building of Agricultural Structures. The

Agricultural building and Structure Engineers are required to build well planned and functional structures. The buildings are needed for meat, poultry, eggs and dairy products, for the storage of grains, corns, maize, vegetable, milk and fruits etc. The Agricultural building and Structure Engineers also make structures like glass houses and plastic tubing for intensive vegetation.

6. Agricultural Electrification - Agricultural Engineers are also involved in electrification. They may also operate electric power system for a rural region. Agricultural Engineers are also in demand for the development or improvement of ways to use electric power for curing drying crops or dehydrating food.

Career Options for Agricultural Engineers

There are good job prospects for Agricultural Engineers. There is a good chance of succeeding in the conservation industry. Here the reuse of agricultural waste material is one of the most important issues. With the ever increasing world population, there is a great demand for farm products. This is undoubtedly a boon for the agricultural industry and especially Agricultural Engineers.

The global trend towards standardization of all agricultural products and equipment would lead to greater demand for agricultural engineers. The earth is going through a crisis which has led to stress on conservation of scare resources. Efforts are being made to achieve high efficiency in the agricultural industry. There is also an increasing demand of agro-products. All this has resulted in multiple opportunities for Agricultural Engineers. There are enough jobs for Agricultural Engineers even after their retirement as they can turn into consultants for big MNC companies.

A large percentage of agricultural engineers work in academia and government agencies and they are also employed for state agricultural extension services. Some agricultural engineers have a penchant for machinery and they may choose the field of designing equipment used in agriculture and construction. The machine manufacturing firms might employ them because of their expertise in crop handling, hydraulic power, crop handling or in growing special crops. Other agricultural engineers may enjoy construction work like designing buildings, storage houses or grains, livestock etc. And some other agricultural engineers might be interesting in developing systems for food processing such as drying processes, distillation or long term storage. Agricultural Engineers work in production, sales, management, research and development or applied science.

There is no doubt that students vying for Agricultural Engineering can expect a bright career ahead.

Salary - Salaries differ significantly with industry, education and experience like for any other engineering branch. The basic salary of an Agricultural Engineer in government institutions is depending upon the seniority of a person. They can expect for a better remuneration in private sectors. If an agricultural engineer decides to open his own farm, there should be no disappointment as farms yield good benefits depending up on the size and dimension of the farm land.

There are also lots of opportunities abroad for Agricultural Engineers. Agricultural Engineers having a bachelor's degree can begin with approx. $43,000 per year. An M.Tech Agricultural Engineer may earn up to $54,000 per year. Those persons with doctoral degree can bring in an annual salary package of $70,000.

This proves that Agricultural Engineering is the most interesting field in modern times. One can enjoy the company of nature as well as be satisfied that a person is creating a healthy with abundant resources for the world and in return gain good income. Indeed agricultural Engineering is a great career option.

Research Activities of Agricultural Engineering

Thrust Area

The Division has presently large number of research schemes financed by state government and Indian Council of Agricultural research. The mechanization of various field operations and popularization of technologies amongst the farmers developed under various project is being done through different extension activities. Large number of technologies developed by the division have been adopted by the farmers. The division is also engaged in fabrication and sale of prototypes under strict quality control for their large scale popularization among the farming community.

To undertake adaptive, applied and basic research in different aspects of Agricultural Engineering for temperate and cold arid zones relating to natural resource development and conservation increasing agricultural production and productivity without drudgery, on-farm post-harvest technology and value addition, transport and management.

To apply engineering principles and practices to rural life with urban amenities improving quality of living, curbing migration to urban areas.

To transfer field worthy technologies to the farmers, craftsmen and manufacturers facilitating agricultural mechanization in the region on-farm agro-processing generating income and employment.

To act as repository of information on aspects of Agricultural Engineering and Technology and their appropriate dissemination for socio-economic development.

Mission

To excel in teaching, research and extension education in agricultural engineering. The division is committed to benefit the state by helping the farming community through mechanization of agriculture and, hence, by increasing the farm production through the conservation and management of natural resources

Transferable Technology Developed

"Design and development of agricultural equipments suitable for the local conditions. More than a dozen of implements and tools for agriculture, horticulture, floriculture, vegetable cultivation etc. have been designed, developed, fabricated, demonstrated, popularized and sold out to the users.

"Protective cultivation in the form of low cost walk in tunnel poly houses was found to be suitable for the temperate climatic conditions existing in the valley. During winter, warm temperatures inside the tunnels promoted plant growth, early flowering and harvesting and better yield, whereas in the open field, plants were dormant. Walk-in tunnel conditions enhanced the early production and harvesting time by 45 days.

"The first fruit under walk in tunnel polyhouse was harvested on 17th March as compared to 2nd May 2006 in open field. The peak harvesting started under polyhouse from 27th March while open field had a peak harvesting on 4th of May.

"The use of mulches also has a profound effect on the yield of strawberry. It was observed that the black mulch was superior over the hay mulch in terms of earliness and yield. The first harvesting of the ripened fruits in black plastic mulch started 2 and 16 days before the hay-mulch polyhouse and open field respectively.

"Irrigation appliances like Sprinkler irrigation and drip irrigation were demonstrated and installed at different places including farmers fields.

"Popularization of power tillers as an alternate source to animal power was done on a large scale. This helped in large scale mechanization in selected regions of the state and also mushroom growth of commercial market for the sale of power tillers.

"In the area of processing of crops, Hop processing plants was installed and managed by the division which helped the hop growers of sumbal and Manasbal and save them from exploitation by middle men and brewers.

"Development and fabrication of harvester for easing the operations of harvesting.Walnut harvesters, apple harvesters , scythes have been fabricated at the division . Demonstrations have also been carried of harvesters procured from different institutions for utilization by the farmers.

"In order to pay the poor farmers the appropriate dues for their produce, a number of graders like walnut grader, tulip grader, saffron grader etc. have been designed, developed and fabricated at the division for use by the poor farmers. All the graders can be manually operated with simple pedaling operation.

"Development of single bullock plough and horse plough has been extensively adopted by the farmers. A number of them have been already sold to the farmers at Wadura, Sopore.

"Walnut crackers like cam operated walnut cracker and impact type walnut cracker have been developed and evaluation and feasibility testing of these implements is in progress.

"Use of hydraulic rams for lifting of water in hilly regions of the state. Over the years a number of them have been installed at different locations of the state like Shalimar, Pattan, K.D. Farm, Ladakh, Kargil etc

"Various bullock operated/ manually operated seed drills for sowing wide variety of crops like wheat, maize mustard, peas etc. have been developed at the division. In addition to this bullock operated Zero- till seed drill, Multi crop seed planter, Mustard seed drill, Ferti seed drill, Pea planter etc. have also been fabricated at the division.

"As paddy is a major crop grown in the region a number of pedal operated/ power tiller operated/motor operated paddy thrashers have been fabricated at the division to ease the operation of thrashing. Which show a considerable saving in time, labour and money in comparison to the traditional methods of thrashing.

"A number of Solar dryers for saffron and walnut have been designed, fabricated and developed at the division the evaluation and feasibility testing of these dryers is in anvil.

"In order to provide other alternatives to cumbersome process of harvesting self propelled reapers and tractor operated reapers have been demonstrated among the farmers for their popularization and large scale use by the farming community.

"Scientific method of growing mat type paddy nursery and its subsequent transplanting by paddy transplanter has been tested and have shown very good results.

21st Century Challenges and Opportunities in Agricultural Engineering

Agricultural engineers made significant contributions to the farming community in the 21st century by improving efficiency of various farming operations, increasing farm incomes, and making life of a farmer more comfortable by mechanizing many of the labor intensive tasks of agricultural production. Mechanization of agriculture is considered to be one of the top 20 technological advancements of the 20th century. With the success of agricultural engineering profession in the 20th century to increase overall global food production, the future challenges of the 21st century still remain not only to feed the current population of 6.5 billion in 2009 but to feed an increasing population of 9.5 billion people in 2050. Despite huge gains in food production, there is still widespread hunger and malnutrition in the world. The problem of world hunger will not be solved until we solve the issue of poor man's access to food and every farmer's access to land, water, credit, and education on this planet. Today we have more than one billion people in poor countries, especially women and children, who go to bed hungry each night and about two billion people are malnourished because of extreme poverty. Therefore, key questions for leaders of agricultural engineering profession would be: i) Are today's agricultural engineers trained to solve global society's top four challenges (energy, water, food and environment) of the 21st century? Ii) Are we providing right education and necessary skills to our agricultural engineering students to face global challenges in 2050? Can we grow enough food to feed 9.5 billion people in 2050? How do we provide clean drinking water, energy, quality healthcare, and clean environment for 9.5 billion people in 2050? Therefore, future of agricultural engineering profession in the 21st century is very bright because large number of agricultural engineers would be needed to solve complex problems of climate change, energy, water, food and environment in developing a sustainable global society in the 21stcentury.

Key Challenges for Agricultural Engineering Profession in 21st Century

With the success of agricultural engineering profession in the 20th century to increase overall global food production, the future challenges of the 21st century would be not only to feed the current population of 6.5 billion and but to feed an increasing population of 9.5 billion people in 2050. Therefore, the major challenge for agricultural engineers would be how to help doubling the food production to feed a growing population of 9.5 billion people on the planet by 2050. Despite the gains made in 20th century in increasing food production, there is still widespread hunger and malnutrition in the developing world. The problem of world hunger will not be solved until we solve the issue of poor man's access to food and every farmer's access to land, water, credit, and education on this planet. Today we have more than one billion people in poor countries, especially women and children, who go to bed hungry each night and about two billion people in the world are malnourished because of extreme poverty. Agricultural engineers must help solve the problems of world hunger and malnutrition. Providing the growing world population with the necessities of life with sustainable solutions will be the prime challenge for agricultural engineers in the 21st century.

Solve Environmental Pollution Problems: Mechanization resulted in developing and adopting intensive agricultural production systems to maximize farm profits. The intensive crop production systems required increased input costs of fertilizer, pesticides and irrigation water to obtain better and better crop yields which resulted in unintended environmental pollution consequences. The environmental problems caused by intensive agriculture systems have been compounding since the World War II period.

Throughout the 20th century, agricultural engineers and plant scientists kept on developing new seed varieties, technologies for fertilizers, pesticides, irrigation application methods, and advanced machines to make production systems automated to increase the crop production year after year. Engineers were successful in optimization the use of irrigation, fertilizers, and pesticides. At the same time, there is a widespread concern about the sustainability of production systems and impact of agricultural production on the water pollution and animal welfare. It was now well accepted and recognized by the farming community and urban society that agricultural production systems take into accounts not only an increase in production but also minimizing the negative impacts on the environment, natural resources and animal welfare.

Biosensors and nanosensors for Food Safety: Public is also now actively engaged in debates on organic farming vs chemically driven farming to assure public health and safety. Therefore, agricultural engineers need to find their role in the overall "sustainable agricultural production systems" of the 21st century and assure public that production systems are sustainable and foods are safe for consumption. Therefore, agricultural engineers have significant opportunities to develop new tools and technologies for risk assessment such as bio and nano sensors for data collection of food safety and traceability. In addition to bringing more land for sustainable food production systems, more and more land is being put under high energy crops to produce either biodiesel or ethanol as new sources of energy. This is another challenge facing the farming community.

Global Positioning Systems for Machines of the Future: Another opportunity for agricultural engineers is blend the knowledge of modern sensors, robots, automation, and manufacturing with information technology and global positioning system (GPS) to develop knowledge based machines of the future. Now robotics and automation are already being widely used in manufacturing and animal production and dairy industry. In the Netherlands, Denmark, USA, Canada, and many other Europeans countries, automatic milking systems (AMSs) are extensively and widely used. The global interest in AMS is very much to save labor cost and maintain better hygiene safety in dairy plants. Another application of sensors and controls would in the animal production facilities where advanced technical solutions are being investigated and used for reducing environmental impact of gases, particles, and odor. Minimizing the use of pesticides for weed control is another emerging field for use the robotics where the crops and weeds can be identified and pesticide sprayers will treat individual areas infected by weeds rather than spraying the pesticide over the entire field. There will be many more emerging new challenges for agricultural engineers and opportunities are going to be unlimited.

Biological Engineering: Another emerging opportunity for agricultural engineers is in the area of Biosystems Engineering. The bioeconomy industry in many countries represents a new generation in the technological revolution that promises to bring high paying jobs to the society. Bioeconomy based industries will need engineers who can transform the knowledge generated by life sciences research into products and services that add value to our agricultural and energy production while protecting the environment. It is envisioned that engineers and scientists from several disciplines will play a critical role in supporting this industry. Historically, agricultural engineers have worked at the interface of engineering and agriculture-based biological systems. In the early 1900's our agricultural engineers focused primarily on development of mechanized equipment, rural electrification and drainage. As our understanding of the biology governing agricultural systems has expanded over the past 25 years, our discipline has evolved, both locally and at the national level, to address more broad-based biological issues related to design and management of agricultural, food processing, and natural resource systems. For example, now agricultural engineers are working as biological engineers in critical areas facing the society including water quality, bacteria and pathogen monitoring and modeling, wetlands and riparian buffer strip design, physiology, bioenergetics, genetics and modeling of plant and animal systems, microbial bioconversion of biomass and waste for nutrients and energy, plant and animal composting, biomaterial harvesting, processing, quality and preservation, food safety and processing, biocontamination and biosensors. Our universities are recommending that we offer an undergraduate degree in biological systems engineering to capitalize on the strength and interest in biological systems engineering and help bioeconomy of grow for future economic systems. Agricultural and biological engineers will have the technical and biological background to understand how intensive agricultural production systems precisely affects the broader ecosystems within a given watershed under varying hydrological systems and what needs to be done to protect ecosystems from agricultural pollution. Intensive agricultural and animal production systems have become source water and air pollution in the developing and developed world future biological engineers must develop new technologies and practices that can help farmers use animal manure as a fertilizer and prevent nutrient losses to air and water resources.

Carbon-neutral Renewable Energy Systems: The future farming systems may be required to limit greenhouse gas emissions and stricter regulations may be in the pipeline for protecting air, soil, and water quality. To remain economically viable, agriculture production systems must be ecologically sound and resource conserved for sustainability. In the future, farmers will realize new revenue from a diverse portfolio of biodiversity, carbon sequestration, water conservation, and renewable energy. Today's agriculture is largely an industrial enterprise sustained and made to appear efficient and economic with large inputs of energy, water, nutrients, and chemicals, derived from nonrenewable natural resources. The availability of inexpensive external energy has allowed agriculture to avoid its own energy production and channel most of its effort into the mass production of food, feed, and fiber commodities. Vast amounts of energy are consumed in machinery and much more energy is expended in producing fertilizers, pesticides, and farm equipment, and in packaging and transporting produce to distant markets. As the true costs of our energy consumption practices become recognized, theneed for a transition to carbon-neutral renewable energy sources becomes more urgent.

In addition to efficient food, feed, and fiber production from agricultural and biological systems, in the future agricultural engineers will work more in the production of ethanol and biodiesel, where crops are grown, harvested, and processed into liquid biofuels.

However, crops and production systems currently available were developed primarily for food, feed, or fiber rather than for energy purposes. Large-scale expansion of biofuels production, therefore, has enormous implications for our agricultural production systems.

Irrigation Systems for Water Scarcity Areas: Better water resource management and stewardship is just one of the many opportunities agricultural and biological engineers will play in the 21st century. Because water is basic to all biological life (human, animal micro-organisms, and plants) the impact of ensuring enough available water for all species will be incalculable. The availability of fresh water supplies for irrigation is declining in intensive irrigated areas of the world. Irrigated agriculture will continue to experience shortages and demands not only for food and fiber but also for emerging demand for supplying energy crops for a worldwide population that is increasing both in size and in industrialization. Future irrigation systems must use less water and provide maximum water application efficiency. Drip, sprinkler, and center pivot irrigation systems are capable of using less water and automatically, efficiently, and uniformly irrigating a variety of crops under different terrains, and soil conditions. Agricultural and biological engineers have improved the safety, efficiency, and dependability of the original design of these irrigation systems but must keeping on improving designs further for efficient use of soil and water, transforming agricultural production throughout the world. With the of precision application technology, irrigation systems will be applying water efficiently while reducing input costs and improving environmental problems in agriculture. To significantly improve irrigation efficiencies will require the use of technologies including remote sensing GPS, GIS-based spatially distributed information, and wireless technology.

The world has won the improtant battle in the area of food security, but the war is still on. A total of 800 million people, that is one out of every six persons, in the developing world do not have access to food. One-third of all pre-school children in the developing countries are food insecure. We, thus, have a big challenge ahead of us as we enter the next millenium.

It is true that the mass starvation that was predicted for Asia in seventies and eighties did not occur. It is only because Science was effectively put to work to raise agricultural productivity. In India, the 'Green Revolution' was a success due to the introduction of improved seeds, fertilizer, irrigation and plant-protection measures combined with positive policy support, liberal public funding for agricultural research and development, and dedicated work of farmers. Notwithstanding all-round achievements, the basic problems of food security, poverty, equity and sustainability, continue to be a cause of concern in India today.

THE INDIAN CHALLENGE

There are several concerns about Indian agriculture that we need to address. Even in the green revolution crops, rice and wheat, the Indian average yield ranks around 50th in the world. A weak agro-based industry, the problems of weak marketing, storage, transportation, credit support, etc, haunt us. Only 0.3% of agricultural GDP is spent on agricultural R &D. The population is growing at an alarming rate of 1.8% per annum. It will be around 1.3 billion by 2020 and would require 50% additional foodgrains. A total of 200 million people are still below the poverty line and have insufficient access to food. Declining resource base, degeneration of natural resources, including soil fertility, water environment, etc, are some other concerns. Food production systems centre around a few states only. Out of 17 big states, only 7 are self-sufficient and surplus in foodgrains. There was a significant decline in per capita intake of calories and protein between 1975 and 1990.

Inspite of all these problems, there is plenty of good news. India harbors a vast diversity of basic natural resources. Tremendous bio-diversity of agriculturally important micro-organisms exists, which are not yet been harnessed. We have a variety of crops, including quality fruits, vegetables, foodgrains, and a number of plantation and commercial crops. Indian fisheries resource has spread over 8,100 km of coastline on the east and west, and provides access to marine resources and transportation.

Very large proportion of area of various food crops is under low productivity category. It can be increased with inputs of high class science. There is a scope to cultivate sizeable portion of waste lands through soil amendment and introduction of agroforestry. Several million hectares of saturated soil can be potentially exploited from rainfed lowlands of eastern India. The India Exclusive Economic Zone (EEZ) of the marine sector, of about 2 million sq.km., has high potential harvestable yields. Only about 20% of the cultivated area is covered under tractor cultivation, great scope of making quantism quantum jump in production and productivity exists, changing consumption patterns offer scope for diversified agriculture. Trade liberalization provides opportunities to reach for outside markets, which were not accessible otherwise. Great gains are possible through agro-based industrialization and overall rural development.

The cultivators and farmers of India have a rich heritage of traditional agricultural wisdom on crop and livestock husbandry and fisheries; based on the principles of the conservation of natural resources and environment. India has an extensive network of extension channels for dissemination of technologies generated by research institutes. India has a strong manufacturing capability and a large network of about 20,000 manufacturers and nearly one million village artisans for indigeneous production of all types of agricultural machinery. Therefore, inspite of several disadvantages, we have a great chance to leap forward, if we enforce the right policy initiatives and implement them.

Biotechnology and India Agriculture

We need to recongnise that, just as the food reequirements of today's population of nearly 6 billion people could not have been met by the technologies of the 1940's, we cannot assume that current practices will feed the population of 8 billion expected by 2020. New approaches are needed in addition to the continued improvement of existing methods of crop and animal husbandry and food proceessing. The strategic integration of biotechnology tools into India agricultural systems can revolutionise Indian farming and usher in a new era. Genetic engineering is clearly the most revolutionary tool to impact agricultural research since the discovery of genetics by Mendel. Prior to genetic engineering, the exchange of DNA mateial was possible only between individual organisms of the same species. With the advent of genetic engineering in 1972, scientists have been able to dientify specific genes associated with desirable traits in one organism. For example, a gene from bacteria, virus or animals may be transfered into plants to produce genetically modified plants having changed characteristics. This method therefore allows mixing of the genetic material from species that cannot otherwise breed naturally.

Genetic engineering can be vital for an agrarian country like India. It can help in minimising the crop damage through disease and pest resistant varieties, reducing the use of chemicals, enhancing stress tolerance in crop plants, thus permitting productive farming on unproductive lands, etc. One could even extend the growing season of crops and minimise losses due to environmental factors on the one hand and increase the shelf life of fruits and vegetables, on the other, thus minimising losses due to food spoilage. We can expand the market vista and improve food quality. Biotechnology can also produce plants that possess healthy fats and oils, possess increased nutritive value, and create a whole range of higher value feeds. Biotechnology has even the power of producing biodegradable plastics, edible vaccines, etc.

It is only through the blending of the 'gene revolution' with our experience in the 'geen revolution' that we can reach our goal of 'evergreen revolution' and also 'nutritional revolution' The advantage of the gene revolution is that it is relatively scale neutral, benefiting big and small farmers alike, It is also environment friendly. Thus, it can be of great help to the smallest farmer with limited resources in increasing farm productivity through the availability of improved but powerful seed. It can also reduce a farmer's dependency on chemical inputs such as pesticides and fertilisers.

Modern biotechnology offers unlimited opportunity for enhancing genetic potential of crops and other commodities, management of biotic and abiotic stresses, bio-remediation and organic recycling. India is one of the main centres of agricultural biodiversity and its gene richness can greatly complement the developments in modern biotechnology. Likewise, new developments in GIS, remote sensing, and crop modelling, provide new opportunities for integrated management of natural resources. The revolution in informatics provides opportunities for sharing latest information for research and planning in a highly organised and efficient manner. With the globalisation of economy, the opportunities for value addition and post harvest management are also immense. With clever blending of technologies, the India farmer can usher in new era of confidence and performance.

Sir Francis Bacon had once said: 'It would be an unsound fancy to expect that things which have never yet been done, can be done except by methods which have never been tried' In the new context, we have to review the new role of biotechnology. Also, it is in this context that we have to view the reapid advances that modern biotechnology is making, including the area of transgenics. Let us focus on some of the developments in this area, since considerable interest and controversy has been created around the world today in this exciting area.

Movement in transgenics

The advent of genetic modification over the last two decades has enabled plant breeders to develop new varieties of crops at a faster rate than was possible using traditional methods, with huge potential for further beneficial development. Tobacco was the first plant to be genetically transformed\, in 1983, with cereals beginning in 1990, but it is only recently that products such as GM soya have reached the market place in Europe, and, for a variety of reasons, given rise to concern & controversy. The concern must be addressed seriously if society is to exploit the new technologies appropriately. We will address this issue rather extensively a little later. Whereas in India, we have been discussing and debating the issues, the rest of the world is galloping ahead. See how the agro biotech product sales, includes transgenic seeds, is increasing around the world (table 1). The global area of cultivation for transgenics increased 4-5 fold from 7 million acres in 1996, to 31.5 million acres in 1997, for just 7 crops in 6 countries. The emphasis has been mainly on traits for herbicide, insect, virus & fungal resistance, and on product quality with a limited number of genes. Out of a total 159 releases. Argentina leads the way with 43, followed by chile, Mexico, Puerto Rico and the Republic of South Africa, with 17-20 each. India is far behind with just a few releases.

There are fears relating to transgenic plants in society. These need to be addressed, not by brushing them aside, but by careful scientific analysis and educational programmes. All biosafety protocols must be followed rigidly. Let us deal with this issue more exhaustively now.

Public Concerns

The potential of biotechnology as a method to enhance agriculture productivity in the future has been accepted globally. However, because of its revolutionary nature, risk and uncertainty may be created by the process of genetic engineering and by the resulting genetically modified products. This may result from the introduction of a new unrelated DNA sequence into a recipient organisam. The introduced DNA may have some unexpected effects on the cellular processes of the recipient organism. In addition to this, introduction of antibiotic resistance genes, as selection markers, pose serious implications in public health especially in genetically modified plants that are directly used for food purposes.

Risks are also associated with the genetically modified plants that are released into the environment. The nature of interactions with other organisms of the natural ecosystems cannot be anticipated without proper scientific testing. For example, modified plants with enhanced resistance to pests or disease threaten to transfer resistance to the wild relatives, which will have implications for biodiversity and ecosystem integrity. These, and many more, doubts plague the minds of common people. Therefore, it is very clear that if the biotechnology potential has to be completely realized, it has to be done in an extremely responsible manner. At each step, proper testing and scientific data have to provided. Not only this, proper regulatory and policy mechanisms have to be put into place. It is therefore, absolutely essential, that conscious efforts be made especially by the developing countries to initiate well-defined programmes for the development and regulations of genetically modified plants. To make this possible and to benefically, ethically and sustainably reap the benefits of biotechnology, there is a pressing need for scientists, researchers, policy makers, NGOS, progressive farmers, industrialists, and representatives of the government, to come together on a common forum in order to discuss the common concerns and find solutions to them.

Will modified plants transfer their introduced genes into wild relatives growing nearby?

Will modified plants that produce new compounds disrupt the 'balance of nature' in some way?

Could the planting of a restricted number of cultivators leave crop plants more susceptible to disease?

Could the planting of a restricted number of cultivators lead to a reduction in biodiversity (of crop plants, weed species, insects and microflora in the fields in question)?

Will genetically modified plants be able to avoid the factors that regulate natural populations and threreby change the usual ' balance' between populations?

Similarly, there are concerns about the social and economic effects. For example, some of these concerns are :

How will the structure of farming (particularly in developing countries) be affected by biotechnology?

How will patent laws affect traditional breeders 'rights (e.g. the right to save seed from one year to the next)?

Will plant bredding be left increasingly in hands of a few companies, and if so, what effects might this have?

Will some countries be 'plundered' for their genetic resources?

We also have a set of ethical and moral issues that need addressing. In particular,

The consumer has the right. What will consumers be told about the new food products?

Is it acceptable to 'interfere' with nature through genetic engineering?

Do we have the right not to use all means available to improve crop plants, especially when so many people are under or malnourished?

Then there are several regulatory issues. Among these, the three key issues are:

Do current regulations give sufficient protection to farmers, consumers, those who have invested in, and those engaged in research?

Is there sufficient international legislation to ensure environmental protection?

Do current regulations compromise the competitiveness of biotechnology companies by being excessively restrictive?

The Royal society of London had appointed a group of experts to examine various aspects including the scientific evidence concerning the risk of transfer of genes from Genetically Modified (GM) crop plants to wild species and non-GM crops, the uptake of genes from GM food by the digestive system, and the current state of the regulatory system. The experts concluded that the chances of gene transfer happening are slight, provided the regulatory processes are followed, but that this must be kept under consideration.

The experts also concluded that the uptake of genes via the food chain is niot a new issue because genes (i.e. DNA) are normal constituents of the human diet. Many products from GM plants, such as sugar prepared from GM tomato paste, are so similar that they are regarded as 'substantially equivalent' others, for example flour from GM soya, may contain a new gene or its product, although many of the purification processes involved in food prodution will destroy and DNA present in the raw material.

Some GM foods have been produced using an antibiotic resistance 'market' gene, which is a laboratory device designed to identify genetically transformed platns. The Royal Society has shown concern about the use of such genes in food products and it suggested that any further increase in the use of such markers in the human ir animal food chain would be undersirable. Gm crop plants have been produced to improve insect tolerance and virus resistance, and to include herbicide tolerance, so that transferred to non-target species of plants, and that the development of resistance by target pests is minimised, the regulatory authorities must be assurd that any negative effects would be no greater than those resulting from conventional procedures; and that any long term effects on the environment and ecology would be clsoely monitored, with statutory restrictions in place to control marketing; and also that best practice advice was adopted by growers.

The reliance on a case by case approach to the legislation, may result in lack of analysis of the overall impact of the technology on agriculture and the environment, and of the long term effects of gmoS. aLTHOUGH MECHANISMS ARE ALREADY IN place to regulate many individual aspects of GM technology, there is no means for looking at GM technology as a whole. As independent, overarching, regulatory body is needed to span departmental responsibilties, monitor the enforcement of existing or future regulations, and strengthen the guidelines to growers of GM crops, such as those specifying the isolation distances between Gm and non-GM crops. The body should also review and monitor the membership of advisory committees and regulatory bodies.

2

MECHANISM OF AGRICULTURE

Mechanized agriculture is the process of using agricultural machinery to mechanize the work of agriculture, massively increasing farm output and farm worker productivity. In modern times, powered machinery has replaced many jobs formerly carried out by men or animals such as oxen, horses andmules.

The history of agriculture contains many examples of tool use, but only in recent time has the high rate of machine use been at such a level.

The first pervasive mechanization of agriculture came with the introduction of the plough, usually powered by animals. It was invented in ancient Mesopotamia.

Current mechanized agriculture includes the use of airplanes, helicopters, trucks and tractors, among other vehicles. Modern farms even sometimes use computers in conjunction with satellite imagery and GPS guidance to increase yields. The need for mechanization is as a result of the global population increase which must be fed. It improves the production efficiency by reducing cost per unit of product,encourages large scale production and improves the quality of farm produce. On the other hand, it displaces unskilled farm labor, causes environmental pollution, deforestation and erosion.

HISTORY

Jethro Tull's seed drill (ca. 1701) was a mechanical seed spacing and depth placing device that increased crop yields and saved seed. It was an important factor in the British Agricultural Revolution.

Since the beginning of agriculture threshing was done by hand with a flail, requiring a great deal of labor. The threshing machine, which was invented 1794 but not widely used for several more decades, simplified the operation and allowed the use of animal power.

Before the invention of the grain cradle (ca. 1790) an able bodied laborer could reap about one quarter acre of wheat in a day using a sickle. It was estimated that for each of Cyrus McCormick's horse pulled reapers (ca. 1830s) freed up five men for military service in the U.S. Civil War. Later innovations included raking and binding machines. By 1890 two men and two horses could cut, rake and bind 20 acres of wheat per day.

In the 1880s the reaper and threshing machine were combined into the combine harvester. These machines required large teams of horses or mules to pull.

Steam power was applied to threshing machines in the late 19th century. There were steam engines that moved around on wheels under their own power for supplying temporary power to stationary threshing machines. These were called road engines, and Henry Ford seeing one as a boy was inspired to build an automobile. With internal combustion came the first modern tractors in the early 1900s, becoming more popular after the Fordson tractor (ca. 1917). At first reapers and combine harvesters were pulled by tractors, but in the 1930s self powered combines were developed.

The horse population in the U.S. began to decline in the 1920s after the conversion of agriculture and transportation to internal combustion. Peak tractor sales in the U.S. were around 1950. In addition to saving labor, this freed up much land previously used for supporting draft animals.

The greatest period of growth in agricultural productivity in the U.S. was from the 1940s to the 1970s, during which time agriculture was benefiting from internal combustion powered tractors and harvesting machines, chemical fertilizers and the green revolution.

Agricultural mechanization

Increased agricultural production is most often brought about by the introduction of improved crop varieties and by creating an optimal environment such that the plants and animals can develop to their full potential. Planting, tending and harvesting a crop requires both a significant amount of power and a suitable range of tools and equipment. Mechanization of farming has allowed an increase to the area that can be planted and has contributed towards increased yields, mainly due to the precision with which the crop husbandry tasks can be accomplished. In fact, most farmers in developing countries experience a greater annual expenditure on farm power inputs than on fertilizer, seeds or agrochemicals.

Crop production systems have evolved rapidly over the past century and have resulted in significantly increased yields. Unfortunately, on some occasions the production systems have created undesirable environmental side-effects amongst which soil degradation and erosion, pollution from chemical fertilizers and agrochemicals and a loss of bio-diversity are just a few of the examples that have been highlighted over recent years. Furthermore, not only were some crop production systems found to be unsustainable in an environmental sense, in some locations neither were they sustainable in an economic sense. Of equal concern was the observation that in some cases it was only the work undertaken by men that was mechanized. The tasks traditionally performed by women remained unchanged although the work demanded of them increased as the area planted and the yields increased.

It is against this background that the work in agricultural mechanization has focused on the following aspects:

- All types of farm power (human, animal and mechanical) including the related social, economic and environmental dimensions;
- Standards for farm tools, machinery and equipment, together with codes of conduct for their safe use (implemented in close collaboration with the plant production and protection division);
- Technical, policy and strategy issues concerning mechanization;
- Alternative crop establishment technologies such as conservation agriculture.

Agricultural Mechanization Strategy

Agricultural Mechanization embraces the use of tools, implements and machines for agricultural land development, crop production, harvesting, preparation for storage, storage, and on-farm processing. It includes three main power sources: human, animal, and mechanical. The manufacture, distribution, repair, maintenance, management and utilization of agricultural tools, implements and machines is covered under this discipline with regard as to how to supply mechanization inputs to the farmer in an efficient and effective manner.

Hand tool technology is the simplest and most basic level of agricultural mechanization: the use of tools and simple implements using human muscle as the main power source. Draught animal technology refers to implements and machines utilizing animal muscle as the main power source.

Mechanical power technology is the highest technology level in agricultural mechanization. It embraces all agricultural machinery which obtains its main power from other sources other than muscular power.

Within each of these three levels of mechanization technology, degrees of sophistication must be distinguished. For example, a simple locally made single-axle tractor without differential gears and gear box, a single axle tractor with gearbox and power-take-off, and a 70 kW tractor, are all mechanical power technology, but with a large difference in sophistication and capability.

Terminology such as "intermediate technology" and "selective mechanization" are either inappropriate or have no practical meaning. The term appropriate mechanization may be used, and refers to the level of mechanization and how it is used for a specific situation. "Appropriateness" can only be determined after carefully considering the technical, economic and social characteristics of eachsituation. No generalizations can be made concerning the appropriateness of a particular type of mechanization or particular agricultural tool, implement or machine for rural development. For the purpose of simplicity, the term mechanization in this Section will be used to cover all levels of mechanization technology and their degrees of sophistication.

The purpose of an agricultural mechanization strategy (AMS) is to create a policy, institutional and market environment in which farmers and other end-users have the choice of farm power and equipment suited to their needs within a sustainable delivery and support system. "Farmers and others" refer to all end-users of farm power, tools and equipment, such as small family operated farms, commercial farm businesses, farmers organizations, irrigation groups, contractors, government operators and primary agricultural produce processors (FAO 1997). An AMS deals with manual, draft animal, and mechanical power, the utilization of tools, implements, machinery, their supply and maintenance. The strategy may cover importation and domestic manufacture of tools, equipment and machinery, their repair and maintenance, relevant training and extension programs, improvement of draft animal health services and breeding program, and promotion of financing systems for the purchase of draft animals and machinery.

In the past agricultural mechanization in developing countries has been much criticized because it often failed to be effective, and was blamed for exacerbating rural unemployment and causing other adverse social effects. This was largely the result from experiences during the 1960s until he early 1980s when large quantities of tractors were supplied to developing countries either as a gift from donors, or on very advantageous loan terms. In particular projects which were designed to provide tractor services through government agencies have a miserable record. These projects proved not sustainable because of the intrinsic inefficiencies of government-run businesses. An overvalued foreign exchange rate and low real interest rates made agricultural machinery artificially cheap as compared with labor and draft animals.

These experiences, often combined with a very narrow perception and lack of knowledge about mechanization, namely the one sided promotion of tractors and other capital-intensive mechanical power technology, has caused the aid community to largely turn its back on mechanization. At the same time there are many examples were mechanization has been very successful, contributing to increased food production, productivity and advancement of rural economies. For example, privately owned shallow tube wells for irrigation in South Asia, axial flow threshers in Southeast Asia, single-axle tractors in Thailand, and various forms of farm mechanization in many parts of China.

The introduction of agricultural technology, including mechanization, is a complex process. The formulation of an AMS requires comprehensive knowledge of many aspects of agriculture in its broadest sense. An AMS will very much depend on country specific characteristics of the economy, its level of development, and the agriculture sector. This means that the formulation process for an AMS cannot be prescribed in a simple set of guidelines. Therefore, the purpose of this Section is to aim at a better understanding of the process of mechanization, to provide broad guidelines for strategy formulation, and to address the major issues involved. For those who will be involved in the formulation of an AMS, it is recommended that they also read FAO's guide for AMS preparation (FAO 1997).

The requirements for an AMS must be kept in the proper context. AMS formulation should not become an overbearing time consuming exercise, absorbing excess manpower and other resources, as compared to the actual constraints which need to be resolved by agricultural engineering. Preferably mechanization technology should be considered in the context of an overall (agriculture) technology strategy.

The Need for Mechanization / Productivity Enhancing Technology.

Farm Power -- consisting of manual labor, agricultural tools, draught animals, tractors, implements, equipment, and machinery -- is an essential farm input. In almost any agricultural production system the annual expenditure on farm power, whether on labor, draft animals, or fuel and depreciation of machines, largely exceeds the costs of other inputs such as agro-chemicals and seeds. In many developing countries, agricultural production and food security are adversely affected because of insufficient use of farm power, low labor productivity and/or labor scarcity. The need to improve agricultural labor productivity is increasingly recognized. In the case such as pump sets for irrigation, the need for machinery is undisputed. Rather than agricultural mechanization, it would be preferable to use the term Farm Power or Labor Productivity Enhancing Technology, to recognize not only the importance of manual labor and handtools, draft animals, and mechanical power, but also other issues related to labor scarcity, such as cropping- and farming systems. Finding solutions to environmental problems in agriculture requires (improved) agricultural tools and machinery, for example for soil tillage and pesticide application, the latter also addressing health concerns. Similarly, machines are required to assist with post-harvest loss reduction and on-farm processing. Thus it is now (again) recognized that agricultural mechanization is crucial in the fight against hunger and poverty, and at the same time to address environmental and health concerns. In order to avoid recurrence of the past mistakes such as described in the introduction, formulation of efficient mechanization strategies are required.

When reviewing the process of applying labor-saving (or labor productivity enhancing) innovations in agriculture, it is a serious but frequently made mistake to assume that this can be achieved only through applying mechanical engineering technology. In this context, nine different stages in the process of enhancing labor productivity may be distinguished (Rijk 1989):

Stage I: Application of improved Hand tool Technology. This process started in prehistoric times when early civilizations developed stick and stone tools which were the only means to enhance labor productivity. In many parts of the world, handtools are the only technology used in agriculture, and even in highly mechanized agricultural systems, improved handtools are still important.

Stage II: Draft animal power application. At this stage animal muscle power is substituted for human power, a process which already started in ancient civilizations. A large variety of implements and machines have been developed which use animals as the principle power source.

Stage III: Stationary Power Substitution. Mechanical power is substituted for human and animal power, used in stationary operations. Stationary operations are mechanized first because motive power sources required to move across the field are technically more complex and therefore require higher investment. Typically, operations mechanized at this stage are paddy dehusking, grain milling, pumping water, and threshing.

Stage IV: Motive Power Substitution. At this stage, substitution of mechanical power for muscle power takes place for field operations. It focuses on power-intensive field operations (for example, plowing), and machinery is of relatively simple design, and easy to operate. Mechanization is still straightforward, and crop production practices are usually unchanged.

At Stage III and IV, mechanization takes advantage of lower costs of new power sources as compared with traditional ones.

Stage V: Human Control Substitution. At this stage the emphasis is on substitution of the human control functions. Depending on the complexity of the control function and the degree of its mechanization, machinery becomes increasingly complicated and costly. A potato lifter is simple in design, but fruit and cotton harvesting machinery are complex and expensive .

Stage VI: Adaptation of Cropping Practices. This stage features the adaptation of the cropping system to the machine. For example, removing weeds in broadcast crops cannot be done with machines but row seeding and seed drills may be introduced to facilitate mechanization of weeding. Other examples include the increase in row distance to accommodate heavier and larger machinery to speed up field operations.

Stage VII: Farming System Adaptation. The farming system and production environment is changed to facilitate further increase in labor productivity and to benefit from economies of scale, necessary to make the investment in expensive machinery financially feasible. An example of this is the rapid decline of mixed farming systems in Europe since the late 1960s when farmers specialized in either dairy, poultry, hog, or crop production. Some crops which are difficult to mechanize may disappear if acceptable substitutes become available, or if these can be produced in countries with low labor costs. At this stage, investments in land development, land consolidation, and rural infrastructure are often needed to facilitate advanced degrees of mechanization.

Stage VIII: Plant Adaptation. This stage features the adaptation of the plant and animal to the mechanization system. Mechanization has advanced to a stage were engineering alone can no longer provide further gains in labor productivity. Breeders increasingly take into account the suitability of new varieties for mechanized production.

Stage IX: Automation of Agricultural Production. This stage is progressing in countries with high labor costs and sophisticated demands on production and quality. Examples are automated rationing of concentrate feeding for individual dairy cows based on their milk production, and sprinkler irrigation systems activated by soil moisture.

Basic Guidelines and Principles for Strategy Formulation

The appropriate choice and subsequent proper use of mechanized inputs into agriculture have a significant effect on agricultural production and productivity, the profitability of farming, and on the environment. In most cases, the mere application of advanced tools, draft animals or machines does not by itself lead to increased yields, but is usually applied to reduce cost of production and to counteract peak periods with labor shortage. However, the benefits achievable by using advanced and improved inputs such as irrigation, better seed, fertilizer and pesticides cannot be fully realized without an increased application of farm power. In situations where land is not a constraint and area expansion is feasible, increases in farm power input have lead to direct increase in production by simply increasing the land area or the number of animals that one person can handle.

In agricultural production, farm power is only one input like any other input such as land, irrigation, fertilizer, seed and crop chemicals. The level of its use is one of a mix of management decisions a farmer has to make in order to maximize production and income. Therefore, in a free market situation, it is inappropriate for governments to have a stand-alone policy on mechanization, except as a component contributing towards the realization of a broader agricultural and socio-economic policy. To have a policy to mechanize or, for example, to replace draft animals with tractors, would imply that mechanization is an end in itself, whereas it should only be a strategy to maintain or increase agricultural production or labor productivity.

Some basic principles must be kept in mind in the discussion on AMS:

1. History has shown that the process of mechanization is driven by changes in relative prices, in particular cost of labor versus cost of capital.
2. The reasons for mechanization are economic. The driving force behind mechanization is the farmer's effort to increase or maintain net income.
3. Mechanization is demand-driven: ultimately it is the farmer who will decide what machine to buy, from whom to buy, and how to use it.

Migration: A Mechanism to Manage Agricultural Risk

As we increasingly recognize the far reaching impacts of climate change, research has turned greater attention to the concept of environmentally-induced migration. Vulnerable populations such as the poor and landless are more likely to migrate as a result of environmental shocks, such as drought or excessive rainfall. Recent studies of climate-related migration in northern Nigeria and rural Ethiopian highlands enhance understanding of additional factors that interact with the environment to affect population mobility.

Due to a lack of formal institutions to reduce household vulnerability in Nigeria, internal migration is a strategy commonly employed to diversify income sources, which mitigates agricultural risk. Because weather-related variability such as rainfall and temperature fluctuation will rise with climate change, intensifying agricultural risk has led to a concern about a potential rural exodus into urban areas, which would strain limited resources in Nigerian cities.

Panel data examining migration patterns in Ethiopia reveal that during periods of drought men are more likely to migrate, while women are more likely to remain at home. Data drawn from the Ethiopia Rural Household Survey demonstrate that men from households that own little or no land have especially high labor mobility during times of drought, indicating that poor and landless populations are more vulnerable to environment-induced migration. Interestingly, women are less mobile in response to environmental shocks, which negatively impact wedding finances and heighten the value of female labor in the home.

Understanding how migration is used as a strategy to manage risk to agriculture can illuminate the various social and economic dimensions in a household's ability to adapt to weather-related shocks. This will inform the design of public services and resource allocation in order to best respond to policy challenges arising from climate change and trends towards urbanization.

Agricultural Machinery Sector in India

Over the last few years, there has been considerable progress in agriculture mechanization. It is generally believed that the benefits of modern technology have been restricted to farmers with large land-holdings. Yet the fact remains that even small farmers are adopting and utilising selected farm equipments for efficient farm management through custom hiring. Mechanical equipments for various farm operations like tillage, sowing, irrigation, plant protection and threshing etc are generally being used by the farming community.

Moreover, reliance on animate power for day to day management of farm operations is showing a continuous decline owing to higher gains in form of improved land and labour productivity resulting from use of mechanical power. Improved implements such as M B plough, puddler, disc harrow, peg tooth harrow, spring line harrow being more efficient have been adopted. Further, use of sowing or planting devices is also registering a higher growth due to their impact on seed and fertiliser use. The number of draught animals has also shown a decline as a consequence of farm mechanization.

India in the Global Farm Equipment Market

Indian market is expected to contribute nearly 10 % in 2006 to the worldwide farm equipment market estimated at $ 66 Billion.

Over the last few years Indian agricultural equipment market has been widely exposed to international trade.

Global majors like John Deere, New Holland, Carraro and Same Deutz have already made an entry in Indian market, whereas Indian players are looking for alliances with foreign partners overseas.

Many factors affecting agricultural equipment sales in India include the monsoon, government declared support prices for crops, commodity prices, crop production expenses (including fuel, fertilizer, pesticides and other costs) and the credit policy announced by banks.

Status of Farm Mechanization in India

Even though farm mechanisation shows an increasing trend, there are wide ranging disparities in the levels of mechanisation across states.

Northern States such as Punjab, Haryana, Uttar Pradesh (particularly Western and Tarai belt) have achieved a faster growth in mechanization over various Plans

The sale of other implements and machines like combine harvesters, threshers and other power-operated equipment have been increasing almost throughout the country The pace of mechanization in North-Eastern States has not been satisfactory due to constraints such as hilly topography, socio-economic conditions, high cost of transport, lack of institutional financing and lack of farm machinery manufacturing industries

Mechanization in Western and Southern states of the country viz., Gujarat, Maharashtra, Rajasthan and certain areas of Tamil Nadu, Andhra Pradesh etc., has increased with the increase in area under irrigation and also with the growing awareness among farmer.

The Overall Indian Farm Equipment Market

The market for farm equipments in India is on a growth phase post a decline period that lasted during period 1999 to 2002. The reasons behind the revival include easy availability of finance schemes and reduction in interest rates. The Government focus on building and improving infrastructure also had a positive effect on the overall performance of the farm equipment sector. Contract farming, which is being encouraged by the Government since 2005, will also have a favourable impact on the farm equipment demand. Farmers will now be hedge their crop risks through technical advice provided to them by the corporate organisations, besides assurance of upfront prices and market outlets would increase the farmer affinity for mechanized Agri technologies.

Agricultural Machinery Marketing and after sale-services

The large and medium scale manufacturers have well organized distributors and dealers through out the country to undertake advertising and product promotion in their respective territories, conduct product awareness training programmes for the prospective customers, provide after-sales-service to the customers including free services, repair and maintenance, supply of parts, etc. Therefore, this organized sector has the whole of the country as their market due to which their production volumes are large, and their information feed back about their product performance, improvements required in design, production processing or quality, and the new requirements of the farmers to undertake product developments. Very few small-scale industries have established their marketing network and therefore provide service support in their premises.

In the absence of standardization of parts and components farmers are compelled to carry their machines to the manufactures for repair and replacement of parts and components. Due to this, their market size islimited to their proximity, and they are not able to develop their businesses. The village rtisans on the other hand are located in the villages and therefore provide immediate attention to the needs of the farmers in their immediate neighbourhoods. Therefore, the tools and implements, etc. made by them are against specific requirements of individual customers.

Popularization of agricultural machinery

The assimilation of R & D requires an effective technological infrastructure of institutions and services to develop and test prototypes, to set up pilot plants for intensive evaluation and extensive demonstrations besides, training and credit support. New technology also requires network for transfer of technology to the manufactures. Popularization of agricultural machinery in the country is undertaken by the Provincial Governments through Department of Agriculture or Department of Agricultural Engineering. The activities are coordinated by the Department of Agriculture in Cooperation with the Ministry of Agriculture, Government of India. The Ministry of Food Processing promotes technology related to agro-processing.

The extension system deals with the first-line extension projects with a view to: i) demonstrating the latest technologies to the farmers as well as the extension agencies; (ii) testing and verifying the technologies on the farmers field (iii) providing opportunities to get firsthand scientific feed-back; (iv) developing extension or technological models for the state extension systems; (v) providing training and communication support; and (vi) promoting research in transfer of technologies.

Agricultural Mechanization

Many advancements in farming techniques and tools have been manifested since agriculture's beginnings thousands of years ago. The greatest strides have occurred in the last three hundred years. A substantial contribution to Oklahoma agriculture has been the escalation from manual and stock-animal labor to steam- and then gas-powered implements. Although steel plows, mowers, mechanical reapers, seed drills, and threshers contributed to the development of agriculture in the Great Plains and the West, tractors enabled the western farmer to sow and harvest large acreages with less manpower.

Steam-powered equipment and tractors were available in time for the white settlement of Oklahoma Territory in 1889, but most Oklahoma farmers did not invest, at least individually, in the machinery until World War I. Many large farmers used their outdated tools and horse power for everyday farm work and only depended on steam-powered and, later, gas-powered equipment and harvesting crews to bring in the crop. Entrepreneurs would buy combines and other equipment, hire themselves out to farmers, and follow the harvest from south to north. Many wheat-threshing crews began in southwestern Oklahoma and followed an annual path along the plains to Canada. This ritual continued throughout the twentieth century. Another option for early-day farmers in a region was to invest in self-powered equipment as a group and have a large community harvest. The Great Plains wheat farmer did not lose his distrust of combines and tractors and depart from his trusted headers, binders, and threshers until grain prices rose and labor became scarce during the "war to end all wars," World War I. Agricultural historian R. Douglas Hurt asserts that only 30 percent of American farmers owned a tractor in 1945 and that tractors did not outnumber draft animals until 1955.

By 1920 only 3 percent of Oklahoma farmers had obtained tractors. Nationally, the wheat and grain growers of the Great Plains needed this large equipment more than did the cash-crop cotton, sugar cane, and tobacco growers of the South. In Dixie most farms were too small to make mechanization feasible, and machinery capable of harvesting cotton and tobacco did not develop until the mid-twentieth century. Large farms in the South had access to an abundance of cheap labor, and the tenant/sharecropping system was widespread, reducing the need to invest in automated implements

Mapping the geographical distribution of tractors and major crops in Oklahoma's counties in the 1920s and 1930s draws out a general pattern. Two regions emerge, divided by a diagonal line from northeast to southwest Oklahoma, or from just below Tulsa County through Cleveland County and below Comanche County. South and southeast of the line, the soil and topography were suitable for growing cotton. There, most residents participated in Southern culture. A larger population, compared to the total acreage under cultivation, provided more manpower. Farms were smaller, and a tenant-based system existed by the twentieth century. In the counties northwest of the line, the opposite prevailed: soil and topography suited wheat, the people were generally Midwesterners, most land was worked by owners, the population was sparse and scattered, compared to the total acreage under cultivation, and farms were larger.

Viewing the early distribution of tractors against this backdrop reveals that the northwest adopted agricultural mechanization earlier and in larger measure. For example, in 1928 Grant County had as many tractors (1,296) as all of the counties southeast of the dividing line. Eleven years later, a 1939 survey showed a 34 percent increase in tractors for Grant County (1,969). However, the survey also revealed a very large increase in most southern counties, where mechanization had recently been adopted. Love County, for instance, had eight tractors in 1928 and 150 in 1939. The increase was repeated in most southeastern counties, with only two exceptions, Pushmataha (4 in 1928 to 2 in 1939) and Latimer

The overall pattern shows a steady but small increase in northern and northwestern counties, but a relatively large jump in the numbers of tractors in southern counties from 1928 to 1939. The most dramatic change came in Tillman (363 to 1,485) and Kiowa (591 to 1,859) counties. Statewide, the total number of registered tractors in 1928 was 18,260, and by 1939 the number had risen to 46,800. By 1944 the number had grown to 64,835, with the largest percentage of gain again coming in the southern counties. The greater percentage increase in southeastern counties is exemplified by the statistics for Le Flore (29 to 178), Pittsburgh (88 to 314), and Haskell (53 to 170). The change to mechanized equipment may be the result of farm consolidation and "tractoring" tenant farmers off the land during the Great Depression or because of the educational efforts of agricultural extension agents before and during World War II.

After World War I the Agricultural Extension Service strongly advocated the purchase of tractors and farm implements, causing farmers to accumulate loans. The debt of many farmers increased after agricultural prices dropped and markets constricted. This led to increased tenancy and a poorer rural population. By the early 1930s agricultural prices had dropped more than 60 percent, while industry prices for implements and gas-powered machinery had dropped less than 20 percent. With the failure of many farmers, the size of farms increased as holdings were consolidated. New Deal incentives (plowing under crops to reduce the surpluses in the market and a program to increase the ability of farmers to buy tractors) benefitted the large farmer and crippled the already-suffering small farmers, tenants, and sharecroppers. Many of these became the stream of Okies, Arkies, and other refugees to the California fruit fields.

Over time, at least six Oklahoma companies manufactured tractors. These firms included the Farm Engineering Company in Sand Springs, the Oklahoma Auto Manufacturing Company in Muskogee, the Oklahoma Motor Plow Company in Muskogee, the Patch Brothers Tractor Company in Sand Springs, the Western Tractor Company in Tulsa, and the Wood, Knight, Hawk Plow Company in Oklahoma City. None of these enterprises lasted into the late twentieth century.

Another consequence of the mechanization of agriculture is the preponderance of truck farming from the 1920s and afterward. The truck's increasing popularity, coupled with the development of better roads, led farmers to sell their own produce in farmer's markets and along highways. The mechanical spindle pickers that became popular with Oklahoma cotton producers after World War II eased the reliance on workers and beginning in the 1950s imperiled another labor force. The use of airplanes to spread pesticides indicated that most technology could be modified and applied to the farm. The air-conditioned tractor cab is an example. Although mechanization has increased the output of agriculturalists and lessened the hard labor involved, it has also hastened the decline in the number of farmers and contributed to the increase of large and corporate farms in Oklahoma.

In 1994, after the commercial management of China's agriculture insurance, domestic scholars make more analysis about the forming factors of agriculture insurance, they mainly discuss the forming mechanism of agriculture insurance through functioning mechanism of insurance industry's forming factors. In 2004, since the agriculture insurance policy management, discussion made by domestic scholars on the formation mechanism of agriculture insurance is more focused on government subsidies, while it is less mentioned the regional particularity of the formation mechanism of agriculture insurance.

Although many scholars have noticed the formation mechanism of regional agriculture insurance, they discuss the problem more from the provincial or regional scope. This paper applies the basic theory of regional economics and the agriculture insurance, basing on the special properties of agriculture insurance development and from the perspective of the national region scope as well as supply and demand relation to make a systematically theoretical explanation on the regional formation mechanism of agriculture insurance. The studying perspective and the theory explanation of region has expanded the studying scope of agriculture insurance, thus the conclusion makes our knowledge of agriculture insurance and the measures of policy promoting more realistic.

The formation mechanism of regional agriculture insurance refers to the functioning status of the agricultural insurance market formed by the inner factors influencing the forming and operating of agriculture insurance market and the comprehensive functions. Generally speaking, the agriculture insurance depends on the forming of effective demand and supply under the organic combination of market and government as far as the forming mechanism is concerned. While in the forming and functioning process of regional agriculture insurance market, the various factors among regions which influence the effective supply and demand of agriculture insurance vary to different degrees, so that the forming mechanism of agriculture insurance brings different forms in different regions, which is the implementing mechanism and development mode of agriculture insurance present diversity.

The Demand Analysis of Regional Agriculture Insurance

The demand of regional agriculture insurance refers to the total amount of agriculture insurance goods that the consumers of a region are willing and able to purchase in a certain period (individual farmers or agricultural production units) are willing and meanwhile able to purchase agriculture insurance and the amount of goods. Regional agriculture insurance demand = regional insurance needs + regional insurance desire + regional capacity to pay + regional insurance premium + local government behavior+...+X.

The regional analysis of agricultural risk

Because of the different natural and geographical conditions among regions, the types and risk loss displayed by the agricultural risk among regions vary greatly, and especially the natural risk and its loss degree have distinct regional features. China has a vast area of agricultural production, major disasters in different regions vary, and the risk types, the frequency and the intensity, etc. also vary greatly. The high-altitude regions are cold with short frost-free periods, so the crops are vulnerable for freezing; in the middle and lower reaches of Yangtze River and the Yellow River, the terrain is low, so the crops are vulnerable for the flooding disaster; in the northwestern Loess Plateau, there is little rain and drought disaster often strikes; the coastal areas are vulnerable for typhoon, and red tide, etc. The risk types which are determined by the law of geographical and climatic distribution in different regions determine different agriculture insurances in different regions. Meanwhile, different loss properties of various regions and the risk types and different anti-disaster abilities among people have determined the great differences of the insurance premium demand and compensation standard of agriculture insurance.

In the regions where the anti-disaster ability is weak and the economy is underdeveloped, people have strong desire for agriculture insurance, but the purchasing power is inadequate, so it is appropriate with high government subsidies and low premium; Meanwhile, because the direct economic loss is relatively large, the compensation standard should not be set according to the absolute value of the direct economic loss, but rather raise the standard of payment properly. In the regions where the anti-disaster capacity is at moderate level, the economy is underdeveloped, the direct economic loss rate is at moderate level and the desire for agriculture insurance is also at moderate level, but people's purchasing power is relatively strong, the government subsidies, premium rate and compensation standard should be set at moderate level. In the regions where the anti-disaster capacity is relatively strong, the economyis developed, the direct economic loss rate of agricultural risk is low, the desire for the agriculture insurance is low, but people have stronger purchasing pThe regional analysis of economic and social development level ower, the government should take measures of compulsory insurance, low subsidies, high premium rate and low payment standard.

The regional analysis of economic and social development level

The development level of economy and society is an important factor influencing the development of the agriculture insurance. The development level of China's economy and society presents huge differences among the eastern area, central area, western area, northeastern area and different parts in those areas. The regional difference of economic development level will definitely lead to regional diversity of the demand for agriculture insurance. The difference in different regional economic development level, the major risk types, the risk loss degree and the demand degree for the creation of insurance goods are diverse, even though the agricultural risk has the property of universality.

The east area in China is economically developed, the level of marketization is high, and the scope of economic activity is wide, the agricultural production mainly face the natural risk, the market risk and the technical risk, so insurance consumers put forward higher requirement for the output insurance and price insurance of agricultural goods together with the value insurance of agricultural production equipment. And these risks also influence the index of agricultural product variation, rate, and establishment of the compensation standard, which become an important but complicated problem for the development of agriculture insurance in this region, add difficulties to the work of agriculture insurance technology and claims, then, promote creation of agriculture insurance products and the development speed of agriculture insurance. The western area in China is less-developed, the level of marketization is very low, the scope of economic activity is narrow, and the agriculture is basically on the traditional development stage, so it mainly faces the natural risk and the insurance consumers put forward more requirements for the real insurance loss of agricultural goods, thus the frequency of natural risk and the establishment of compensation standard in this region become the major technological problem for local agriculture insurance producers, while there is little difficulty in technicality needed and in the compensation work; but the natural risk is variety, high frequency , and easily affected by seasons, so the amount of work for agriculture insurance is huge, and there is higher requirements for the quality of insurance service. And China's central and northeastern areas which are called "the country barn" are medium-leveled in economic development between eastern area and western area. In these two areas, the natural risk, market risk and technical risk often function comprehensively and there is strong desire for the creation of agriculture insurance.

The regional analysis on the income level of rural residents

The income level of rural residents is positively related to the demand for agriculture insurance, and the regional differences in income level of rural residents determine the regional variation of demands. The regional differences or gaps in income level of Chinese rural residents determine the variation of demands coming from rural residents in different economic zones. Namely, the rural residents in the east have strong ability to pay for agriculture insurance, and have strong consciousness of risk prevention as well. Whether the combination of them can be transformed into strong demand for agriculture insurance, it depends on the percentage of farming income in household income. The higher the percentage is, the stronger the demand is. The paying capacity for agriculture insurance of rural residents from central regions is relatively insufficient, but the agricultural income accounts for a large percentage of family income, therefore the demand for agriculture insurance is on the rise. As for the western regions, the payment capacity for agriculture insurance of rural residents is at a low level. What's more, in some degree, farming income is the only source of family income. Hence, people there have strong desire for agriculture insurance, but the demand is insufficient.

Regional analysis of rural people's insurance awareness

Regional insurance awareness is affected by history, culture, society, and psychology factors in different regions. The difference of history, culture, society and psychology lead to the different insurance awareness in different regions which causes the different demands for insurance.

In the east part of China, insurance business started early and develops quickly. Thus there are many types of insurance products and the insurance awareness of the people there is strong. The mid-west area insurance business starts late, develops slowly, the products are few, and people's insurance awareness is relatively weak. While in these poor remote regions, people did not even know what the insurance is, not mention the modern insurance awareness. From social and the psychological aspect, the opening degree in eastern area is higher and populations there are highly concentrated and floating, thus people in this area are easy to accept new things which are helpful to build up insurance awareness. The mid-west areas develop slowly and people there live in a relatively large area where is not good for building up insurance awareness. In the poor remote and old revolutionary base areas, there are small populations with low speed of floating, and people there live in uninformed and nasty environment, which lead to weak insurance awareness there, even some misunderstanding to insurance.

The Regional Supply Analysis of Agriculture Insurance

Regional agriculture insurance supply refers to, in the condition of a certain insurance price, the total amount of insurance products that all the insurance companies in a region are willing and able to provide. The demand of agriculture insurance is the most vital element for the supply of the regional agriculture insurance in region. In addition, other elements that affect and restrict the agriculture insurance supply include the capital value of manufactures, the paying capacity of consumers, the numbers and quality of staffs related to the regional insurance, the category and price of regional insurance, and regional government behaviors and so on. It can be showed in a formula, the supply of regional agriculture insurance = the capital value of manufactures + the paying capacity of consumers + the numbers and quality of staffs related to the regional insurance + the price of regional insurance + the management capability of regional insurance companies + regional government behaviors + … + X.

The regional analysis of the development of banking and insurance business

The development of regional economy is the premise of the regional agriculture insurance's development and it also determines the development of regional banking and insurance business, the financial power and the scale of financially supporting agriculture of this region, and then it affects the supply and inventive capacity of the agriculture insurance products, in which the development of regional banking and insurance is the main factor influencing the supply of regional insurance products. While the development of the regional banking and insurance business is determined by the economic and social development of the region, which means that the higher the region's economy, the market, social capital, the modern civilization and the banking and insurance business develop, the supply capacity of its products develop. The differences in the level of banking and insurance's development will bring in the difference of agriculture insurance base and the difference of supply capacity of agriculture insurance products.

Since the reform and opening, China's banking and insurance have developed rapidly, the financial volume and structure are changing constantly, and state-owned commercial banks, non-state-owned commercial banks, and foreign-invested financial institutions also developed fast, and relative financial ratios, the credit support ratio of economic growth in all regions are on the rise. However, the eastern area is always on the top of that; while in the middle and west of China, they were equally matched before 1990, but 1990 was a turning point, and the Western financial development obviously exceeded the middle. According to "financial market developing index" that exactly show the development of banking business, both in stock market and insurance market, all the big economic areas have an fast increasing tendency, but the gap is even large among these areas. The financing and premium income in the East is as more than twice as that in the middle and the West, and the middle area is twice the west, which means there is not only absolute financial difference among regions, but the relative difference.

The large differences among regions determine the differences in the base of insurance development and the supply capacity of agriculture insurance products. So we know the base of insurance development and supply capacity of agriculture insurance products in the east area is the best, then the northeast area, and the middle, and the west is the worst.

The regional analysis of the capability to supply and innovation of agriculture insurance products

The capability to supply and innovation of agriculture insurance products depends on the development of insurance, especially the agriculture insurance business. The development insurance business including agriculture insurance lies in infrastructure, capital, information, technology, management, talents, and salary and so on. Thus, these vital factors could not be balanced in the spatial distribution, so the development level of regional insurance is different.

The unbalance of regional insurance development is reflected by economic unbalance of regions. During past 20 years of restoring in China's insurance business, some Eastern area, such as Beijing, Shanghai, Shenzhen and so on have a lot of advantages to bring investors, capital, technology and labor force here, so the East has faster development in insurance business, forming the "center of regional insurance", for example, Beijing, Shanghai, Shenzhen, while, owing to the lack of technology, talents in Middle and Western insurance business, they lag behind relatively. In the aspect of premium income scale, the number of insurance organizations, insurance agency business, as well as intensity and density of insurance, there are tendencies of gradual decrease from the East to the West. The regional difference of China's insurance development results in the difference in agriculture insurance products and innovations.

3

HARVESTING AND THRESHING

Harvest constitutes a major operation among agricultural activities. Considered for a long time as the last step in production, it must rather be approached as the first one in the postproduction system, because of its influence on subsequent processing and preservation of the products. Harvesting methods differ according to the part of the plant to be used. As regards forage crops, the whole plant is cut, but for underground crops (eg, groundnuts, roots and tubers), the crop is lifted while the soil sticking to it is removed. With cereals, the crop is first cut either as a whole or partially (ears), and then threshed and cleaned to separate the grain from the ears and straw.

In the latter case two main alternatives exist: separate harvesting and threshing, or combined harvesting and threshing. In developing countries the first alternative is generally the most widely applied. Although harvesting and threshing are still frequently done by hand, their mechanization has begun to develop during recent years, especially where the crop is produced not for self-consumption but rather for commercial purpose. Nevertheless, such mechanization has not developed everywhere to the same extent but according to the type of crop concerned, because labour requirements remain high for handling the produce before threshing.

In industrialized countries, attempts have been made since the beginning of the 20th century to devise machines which would both harvest and thresh grain, so as to reduce the labour requirements involved. Combine harvesters ('combines') which can cut, convey, separate and thresh the grain were the product of this development work. They are in widespread use, and have been used already on large grain production schemes in a number of developing countries.

Rice Harvesting and Threshing

(i) Harvesting methods

Manual harvesting

In many countries rice ears are cut by hand. A special knife is frequently used in SouthEast Asia ("ani-ani"), Latin America ("cuchillo") and Africa. For instance, in the Casamance region of Senegal rice is cut stem by stem with a knife, 10 cm below the panicle so as to leave straw in the field in amounts large enough to produce grazing for cattle. Nevertheless such practice is labour intensive.

To harvest denser varieties (500 stems/sq metre instead of 100) a sickle is used mainly on a generally wetter produce. But work times remain high: 100 to 200 man-hours per ha for cutting and stooking.

Mechanized Harvesting

During past decades the mechanization of rice harvesting has rapidly evolved. It first developed in Japan, then in Europe and has now reached many tropical countries.

The first machines used were simple animal-drawn (horses in Europe, oxen in the tropics) or tractor-driven mowing machines fitted with a cutter bar. The improvements made on this equipment have first resulted in the development of swathers. These drop the crop in a continuous windrow to the side of the machine making it easy to pick up the panicles and manually tie them into bundles. The next step forward has been the reaper that forms unbound sheaves; and finally the reaper/binder which has a tying device to produce sheaves bound with a twine. However the supply, cost and quality of the twine are the main problems associated with the use of such equipment.

The output of these machines varies between 4 and 10 hours per hectare, which is slow. However, they may be usefully introduced into tropical rice growing areas, where hand harvesting results in great labour problems. In temperate countries they have been gradually replaced by combine harvesters.

(ii) Threshing Methods

After being harvested paddy bunches may be stacked on the plot. This in-field storage method results in a pre-drying of the rice ears before threshing, the purpose of which is to separate seeds from panicles.

Traditional Threshing

The traditional threshing of rice is generally made by hand: bunches of panicles are beaten against a hard element (eg, a wooden bar, bamboo table or stone) or with a flail. The outputs are 10g to 30kg of grain per man-hour according to the variety of rice and the method applied. Grain losses amount to 1-2%, or up to 4% when threshing is performed excessively late; some unthreshed grains can also be lost around the threshing area.

In many countries in Asia and Africa, and in Madagascar, the crop is threshed by being trodden underfoot (by humans or animals); the output is 30kg to 50kg of grain per manhour. The same method, but using a vehicle (tractor or lorry) is also commonly applied. The vehicle is driven in circles over the paddy bunches as these are thrown on to the threshing area (15m to 20m in diameter around the stack). The output is a few hundred kg per hour. This method results in some losses due to the grain being broken or buried in the earth.

In south-east Asia, total losses induced by traditional harvesting and threshing methods are estimated between 5 and 15%.

Mechanized Threshing

From a historical viewpoint, threshing operations were mechanized earlier than harvesting methods, and were studied throughout the 18th century.

Two main types of stationary threshing machines have been developed.

The machines of Western design are known as 'through-flow' threshers because stalks and ears pass through the machine. They consist of a threshing device with pegs, teeth or loops, and (in more complex models) a cleaning-winnowing mechanism based upon shakers, sieves and centrifugal fan. The capacities of the models from European manufacturers (eg, Alvan Blanch, Vicon, Borga) or tropical countries (Brazil, India, etc.) range from 500 to 2000kg per hour.

In the 70s, IRRI developed an axial flow thresher which has been widely manufactured at local level. Such is the case in Thailand where several thousands of these units have been put into use. They are generally mounted on lorries and belong to contractors working about 500 hours per year.

More recently, a Dutch company (Votex) has developed a small mobile thresher provided with either one or two threshers. The machine has been widely adopted in many rice growing areas. The simple design and work rates of these machines (about 500kg per hour) seem to meet the requirements of rural communities.

The 'hold-on' thresher of Japanese design, is so-called because the bundles are held by a chain conveyor which carries them and presents only the panicles to the threshing cylinder, keeping the straw out. According to the condition of the crop, work rates can range between 300kg and 700kg per hour (Iseki model). The main disadvantage of these machines is their fragility.

(iii) Combined Harvesting and Threshing Methods

Combine-harvesters, as the name implies, combine the actions of reaping and threshing. Either the 'through-flow' or the 'hold-on' principle of threshing may be employed, but the reaping action is basically the same. The main difference is that combine-harvesters of the Western ('through-flow') type are equipped with a wide cutting bar (4-5m) while the working width of the Japanese ('hold-on') units is small (1m). According to the type of machine used, and specially to their working width, capacities range from 2 to 15 hours per hectare.

Such machines are being increasingly used in some tropical countries. In the Senegal river delta region, private contractors or farmers' organizations have recently acquired combine harvesters, mainly of the Western type (Massey Ferguson, Laverda, etc.). So, almost 40% of the Delta surface area is harvested with a pool of about 50 units. Between 200 and 300 hectares of winter rice are mechanically harvested. In this region the popularity of combine harvesters is high despite their poor suitability for some small-sized fields.

In Brazil, several manufacturers have adapted machines to rice growing conditions by substituting tracks for wheels; some machines are simple mobile threshers equipped with cutter bars.

In Thailand, local manufacturers have recently transformed the IRRI thresher into a combine harvester so as to reduce the labour requirement. The unit can harvest 5ha per day and seems to have been rapidly adopted.

(iv) Strippers

Because of their size, conventional harvesters and combine harvesters prove unsuitable for many rice growing areas with small family farm holdings. In response to this problem, research services, during the last ten years, have developed small-sized machines for harvesting the panicles without cutting the straw. Such machines are known as strippers.

In the UK, the Silsoe Research Institute (SRI) has developed a rotor equipped with special teeth for strip-harvesting spikes or panicles. IRRI recently adopted this technology and has developed a 10hp self-propelled 'stripper gatherer' with a capacity of about 0.1ha per hour. However, the harvested grain has to be threshed and cleaned in a separate thresher. Since harvesting unthreshed produce results in frequent stoppages for emptying the machine, this constitutes the main drawback to the progress of the prototype.

In France, CIRAD-SAR has designed and developed a machine which strips panicles from the plants and threshes them in only one pass. The stripper has been specially designed for harvesting paddy rice on small plots. The essential component is a wire looped in line with the direction of movement of the machine, which is mounted on a three-wheeled carriage and powered by a 9hp engine. With a 30 cm working width the stripper capacity is about 1 ha per day.

Maize Harvesting and Threshing

(i) Harvesting methods

Manual harvesting

In village farming systems the crop is often harvested by hand, and cobs are stored in traditional structures. Quite often, the crop is left standing in the field long after the cobs have matured, so that the cobs may lose moisture and store more safely after harvest.

During this period the crop can suffer infestation by moulds and insects and be attacked by birds and rodents. To reduce such risks, an old practice (called "el doblado") is sometimes applied in South and Central America. This involves hand-bending the ears in the standing crop without removing them from the stalks. It helps mainly to prevent rainwater from entering the cobs, and also limits bird attacks; but, because of the high labour requirements involved, the practice is gradually falling into disuse.

Manual harvesting of maize does not require any specific tool; it simply involves removing the cob from the standing stalk. The work time averages 25 to 30 days per ha. Traditionally, maize cobs are commonly stored in their unhusked form. To improve their drying, it is often recommended to remove the husks from the cobs. Maize husking is usually a manual task carried out by groups of women. Some machine manufacturers (e.g. Bourgoin in France) have developed stationary maize huskers, such as the "Tonga" unit.

Mechanized harvesting

The first mechanized harvester to detach ears of maize from the standing stalks, the 'corn snapper', was built in North America in the middle of the 19th century. This was followed by the development of 'corn pickers', which incorporated a mechanism for removing the husks from the harvested ears. The first animal-drawn maize pickers were replaced by tractor-drawn units (1 or 2 rows) and then tractor-mounted units (1 row). Finally came the development of self-propelled units capable of harvesting from up to 4 rows. A specific feature in maize harvesters is the header which leaves the stalks standing as it removes the ears.

The rates of work can vary from 2 hours per hectare with a 3-row self-propelled harvester to 5 hours per hectare with a tractor-drawn or -mounted single row unit. Generally speaking, harvest losses range from 3% to 5%, but they may be up to 10%-15% under adverse conditions. Depending on the situation, a single-row harvester can be employed effectively on up to 20 hectares or more; but the use of a multi-row machine demands several tens of hectares to be economically effective.

Specially designed for harvesting maize as grain, the corn-sheller was initially a cornhusker in which the husking mechanism was replaced by a threshing one (usually of the axial type). Corn-shellers are self-propelled machines of the 3 to 6-row type with capacities of 1 to 2 hours per hectare. The surface areas harvested during a 180-hour campaign range between 100ha (with a 3-row unit) and 200ha (with a 6-row one).

Another alternative consists of equipping a conventional combine with a number of headers corresponding to the machine horsepower. However, although widely used, such a method requires many adjustments to the threshing and cleaning mechanisms.

(ii) Threshing methods

Shelling and threshing

Traditional maize shelling is carried out as a manual operation: maize kernels are separated from the cob by pressing on the grains with the thumbs. According to the operator's ability the work rate is about 10kg per hour. Outputs up to 20kg per hour can be achieved with hand-held tools (wooden or slotted metal cylinders). To increase output, small disk shellers such as those marketed by many manufacturers can be recommended. These are hand-driven or powered machines which commonly require 2 operators to obtain 150kg to 300kg per hour. Another threshing method, sometimes applied in tropical countries, involves putting cobs in bags and beating them with sticks; outputs achieved prove attractive but bags deteriorate rapidly.

Motorized threshing

Nowadays many small maize shellers, equipped with a rotating cylinder of the peg or bar type, are available on the market. Their output ranges between 500 and 2000kg per hour, and they may be driven from a tractor power take off or have their own engine; power requirements vary between 5 and 15hp according to the equipment involved. For instance the French Bourgoin "Bamba" model seems well-suited to rural areas in developing countries because of its simple design, easy handling and versatility (maize, millet sorghum, etc.).

Millet and Sorghum Harvesting and Threshing

(i) Manual Harvesting

In Africa, and especially in the Sudano-Sahelian area, these cereals constitute the staple food in the human diet. They are harvested almost exclusively by hand, with a knife after unroofing or bending the taller stems to reach the spikes. Harvesting and removal from the field takes 10 to 20 days per hectare, according to yields. Harvested ears are stored in traditional granaries while the straw is used as feed for cattle or for other purposes (e.g. thatching).

(ii) Gradual Mechanization of Threshing

Women separate the grain from the ears with a mortar and pestle, as it is needed for consumption or for marketing purpose. The threshed grain is cleaned by tossing it in the air using gourds or shallow baskets.

This traditional method is arduous and slow (10kg per woman-day). Consequently, research has been conducted for some years on how to mechanize it.

The mechanical threshing of sorghum ears does not raise any special problems: conventional grain threshers can be used with some modifications; such as adjustment of the cylinder speed, size of the slots in the cleaning screens, etc. On the other hand, the dense arrangement of spikelets on the rachis and the shape of millet ears (especially pearl millet), make their mechanical threshing excessively difficult.

The first millet and sorghum threshers were developed in Senegal in the 1960-70s: the Siscoma BS 1000 and the Marot DAK II. Giving relatively high outputs (about 1000kg per hour) they have been intended for village farmers' groups, cooperatives or private contractors going from village to village to work on big threshing layouts. The multipurpose "Bamba" thresher, better suited to rural communities, has a capacity of about 300kg per hour. The Senegalese pool of millet and sorghum threshers currently amounts to 120-150 units.

As regards mechanized harvesting at family level, some hand-operated threshers (Champenois) were developed and tested experimentally but they did not prove very successful. CIRAD is currently working on the design of powered millet threshers of low capacities (50 to 100kg per hour).

Grain Cleaning

Threshing operations leave all kinds of trash mixed with the grain; they comprise both vegetable (e.g. foreign seeds or kernels, chaff, stalk, empty grains, etc.) and mineral materials (e.g. earth, stones, sand, metal particles, etc.), and can adversely affect subsequent storage and processing conditions. The cleaning operation aims at removing as much trash as possible from the threshed grain.

The simplest traditional cleaning method is winnowing, which uses the wind to remove light elements from the grain.

(i) Mechanized Cleaning

The most rustic equipment is the winnower: a fan-originated current of air passes through several superposed reciprocating sieves or screens. This type of machine was widely used in the past for on-farm cleaning of seed in Europe. It can be either manually powered or motorised; capacities range from a few hundred kilogrammes to several tonnes per hour.

In Europe, with the use of combine harvesters and the development of centralized gathering, cereal winnowers have been progressively replaced by seed cleaners in the big storage centres. These machines, also equipped with a system of vibrating sieves, are generally capable of very high outputs (several tens of tonnes per hour). In developing countries, mechanizing the cleaning operation at village level has seldom been felt as a necessity, because of the lack of quality standards in grain trading. However, because of the current trend towards privatization of marketing networks, the demand for cleaning machines will probably increase. The local manufacture and popularization of simple and easily portable equipment, such as winnowers or screen graders suited to cereal crops, need to be encouraged. CIRAD/SAR has recently developed cleaning machines of the rotary type with outputs of a few hundred kilogrammes per hour.

Harvest

Harvesting is the process of gathering a ripe crop from the fields. Reaping is the cutting of grain or pulse for harvest, typically using a scythe, sickle, or reaper. On smaller farms with minimal mechanization, harvesting is the most labor-intensive activity of the growing season. On large mechanized farms, harvesting utilizes the most expensive and sophisticated farm machinery, such as the combine harvester. The term “harvesting” in general usage may include immediate postharvest handling, including cleaning, sorting, packing, and cooling.

The completion of harvesting marks the end of the growing season, or the growing cycle for a particular crop, and the social importance of this event makes it the focus of seasonal celebrations such as harvest festivals, found in many religions.

Etymology

“Harvest”, a noun, came from the Old English word hærfest, meaning “autumn” (the season), “harvest-time”, or “August”. (It continues to mean “autumn” in British dialect, and “season of gathering crops” generally.) “The harvest” came to also mean the activity of reaping, gathering, and storing grain and other grown products during the autumn, and also the grain and other grown products themselves. “Harvest” was also verbified: “To harvest” means to reap, gather, and store the harvest (or the crop). People who harvest and equipment that harvests are harvesters; while they do it, they are harvesting.

Crop Failure

Crop failure (also known as harvest failure) is an absent or greatly diminished crop yield relative to expectation, caused by the plants being damaged, killed, or destroyed, or affected in some way that they fail to form edible fruit, seeds, or leaves in their expected abundance.

Crop failures can be caused by catastrophic events such as plant disease outbreaks, heavy rainfall, volcanic eruptions, storms, floods, or drought, or by slow, cumulative effects of soil degradation, too-high soil salinity, erosion, desertification, usually as results of drainage, overdrafting (for irrigation), overfertilization, or overexploitation.

In history, crop failures and subsequent famines have triggered human migration, rural exodus, etc.

The proliferation of industrial monocultures, with their reduction in crop diversity and dependence on heavy use of artificial fertilizers and pesticides, has led to overexploited soils that are nearly incapable of regeneration. Over years, unsustainable farming of land degrades soil fertility and diminishes crop yield. With a steadily growing world population and local overpopulation, even slightly diminishing yields are already the equivalent to a partial harvest failure. Fortunately, fertilizers obviate the need for soil regeneration in the first place, and international trade prevents local crop failures from developing into famines.

Other Uses

Harvesting commonly refers to grain and produce, but also has other uses: fishing and logging are also referred to as harvesting. The term harvest is also used in reference to harvesting grapes for wine. Within the context of irrigation, water harvesting refers to the collection and run-off of rainwater for agricultural or domestic uses. Instead of harvest, the term exploit is also used, as in exploiting fisheries or water resources. Energy harvesting is the process of capturing and storing energy (such as solar power, thermal energy, wind energy, salinity gradients, and kinetic energy) that would otherwise go unexploited. Body harvesting, or cadaver harvesting, is the process of collecting and preparing cadavers for anatomical study. In a similar sense, organ harvesting is the removal of tissues or organs from a donor for purposes of transplanting.

Harvesting or Domestic Harvesting in Canada refers to hunting, fishing, and plant gathering by First Nations, Métis, and Inuit in discussions of aboriginal or treaty rights. For example, in the Gwich'in Comprehensive Land Claim Agreement, "Harvesting means gathering, hunting, trapping or fishing..." Similarly, in the Tlicho Land Claim and Self Government Agreement, "'Harvesting' means, in relation to wildlife, hunting, trapping or fishing and, in relation to plants or trees, gathering or cutting."

Threshing

Threshing is the process of loosening the edible part of cereal grain (or other crop) from the scaly, inedible chaff that surrounds it. It is the step in grain preparation after harvesting and before winnowing, which separates the loosened chaff from the grain. Threshing does not remove the bran from the grain.

Threshing may be done by beating the grain using a flail on a threshing floor. Another traditional method of threshing is to make donkeys or oxen walk in circles on the grain on a hard surface. A modern version of this in some areas is to spread the grain on the surface of a country road so the grain may be threshed by the wheels of passing vehicles.

Hand threshing was laborious, with a bushel of wheat taking about an hour. In the late 18th century, before threshing was mechanized, about one-quarter of agricultural labor was devoted to it.

Industrialization of threshing began in 1786 with the invention of the threshing machine by Scotsman Andrew Meikle. Today, in developed areas, it is now mostly done by machine, usually by a combine harvester, which harvests, threshes, and winnows the grain while it is still in the field.

A threshing bee was traditionally a bee in which local people gathered together to pitch in and get the season's threshing done. Such bees were sometimes festivals or events within larger harvest festivals. Today the original purpose is largely obsolete, but the festival tradition lives on in some modern examples that commemorate the past and include flea markets, hog wrestling, and dances.

4

GRAIN DRYING

Grain drying is process of drying grain to prevent spoilage during storage. The grain drying described in this article is that relating to processes supplementary to the natural ones.

OVERVIEW

Hundreds of millions of tonnes of wheat, corn, soybean, rice and other grains as sorghum, sunflower seeds, rapeseed/canola, barley, oats, etc., are dried in grain dryers. In the main agricultural countries, drying comprises the reduction of moisture from about 17-30%w/w to values between 8 and 15%w/w, depending on the grain. The final moisture content for drying must be adequate for storage. The more oil the grain has, the lower its storage moisture content will be (though its initial moisture for drying will also be lower). Cereals are often dried to 14% w/w, while oilseeds, to 12.5% (soybeans), 8% (sunflower) and 9% (peanuts). Drying is carried out as a requisite for safe storage, in order to inhibit microbial growth. However, low temperatures in storage are also highly recommended to avoid degradative reactions and, especially, the growth of insects and mites. A good maximum storage temperature is about 18 °C.

The largest dryers are normally used "Off-farm", in elevators, and are of the continuous type: Mixed-flow dryers are preferred in Europe, while Cross-flow dryers in the United States. In Argentina, both types are commonly found. Continuous flow dryers may produce up to 100 metric tonnes of dried grain per hour. The depth of grain the air must traverse in continuous dryers range from some 0.15 m in Mixed flow dryers to some 0.30 m in Cross-Flow. Batch dryers are mainly used "On-Farm", particularly in the United States and Europe. They normally consist of a bin, with heated air flowing horizontally from an internal cylinder through an inner perforated metal sheet, then through an annular grain bed, some 0.50 m thick (coaxial with the internal cylinder) in radial direction, and finally across the outer perforated metal sheet, before being discharged to the atmosphere. The usual drying times range from 1 h to 4 h depending on how much water must be removed, type of grain, air temperature and the grain depth. In the United States, continuous counterflow dryers may be found on-farm, adapting a bin to slowly drying grain fed at the top and removed at the bottom of the bin by a sweeping auger.

Grain drying is an active area of manufacturing and research. The performance of a dryer can be simulated with computer programs based on mathematical models that represent the phenomena involved in drying: physics, physical chemistry, thermodynamics and heat and mass transfer. Most recently computer models have been used to predict product quality by achieving a compromise between drying rate, energy consumption, and grain quality. A typical quality parameter in wheat drying is breadmaking quality and germination percentage whose reductions in drying are somewhat related.

Grain Drying Fundamentals

Drying starts at the bottom of the bin, which is the first place air contacts. The dry air is brought up by the fan through a layer of wet grain. Drying happens in a layer of 1 to 2 feet thick, which is called the drying zone. The drying zone moves from the bottom of the bin to the top, and when it reaches the highest layer, the grain is dry. The grain below drying zone is in equilibrium moisture content with drying air, which means it is safe for storage; while the grain above still needs drying. The air is then forced out the bin through exhaust vent.

Allowable Storage Time

Allowable storage time is an estimate of how long the grain needs to be dried before spoilage and maintain grain quality during storage. In grain storage process, fungi or molds are the primary concern. Many other factors, such as insects, rodents, and bacteria, also affect the condition of storage. The lower the grain temperature is, the longer the allowable storage time will be.

Proper Moisture Levels for Safe Storage

It is possible for long period safe storage if grain moisture content is less than 14%, and stored away from insects, rodents and birds. The following figure is the recommended moisture content for safe storage.

Equilibrium Moisture Content

Moisture content in grain is related to the relative humidity and the temperature of the surrounding air. Equilibrium moisture content point is the point when grain no longer losing or gaining water when contacting with drying air. The final moisture content of the grain is up to the amount of moisture in the drying air, which is the relative humidity. The low relative humidity means air is dry and it has a large potential of picking up water. The lower the relative humidity is, the drier the air is. In general, one-half reduce in relative humidity is caused by 20° degree increase in air temperature.

Temperature

Heated air may be used in grain drying process. It can not only accelerate moisture migration inside the kernel, but also can evaporate the moisture on the surface. The major problem about heated air for drying process is the kernel temperature. The grain kernel may be damaged by high kernel temperatures. Usually, kernel temperature is lower than the air temperature. For different use of the corn, temperatures vary. For example, for seed corn, the maximum temperature is 110 °F; for livestock feeding corn, the maximum temperature is 180 °F.

Aeration

Aeration process refers to the process of moving air through grain. Airflow is a measurement of the amount of air in cubic feet per minute (CFM). In grain drying process, drying time is largely depended on aeration rates. Without sufficient airflow, grain may be damaged before drying is complete. Fans are used to move air through grain.

Drying Cost

The drying cost is made up of two parts: the capital cost and the operating cost. Capital cost is largely depend on the drying rate requirement, and equipment cost. Operating cost refers to fuel, electricity and labor force cost. The amount of energy required to dry a bushel of grain is similar for all the drying methods. Some methods depend largely on natural air, while others may use LP heat or natural gas, which make energy cost vary. Basically, fuel and electrical power are the major portions of the operating cost. Drying cost is based on the B.T.U. consumption of temperature change from environment to desired one.

CLASSIFICATION OF GRAIN DRYING METHODS

In-storage drying methods

1. Low-temperature drying
2. Multiple-layer drying

Batch Drying Methods

1. Bin batch drying
2. Column batch drying

Continuous Flow Drying Methods

1. Cross flow drying
2. Counter flow drying
3. Concurrent flow drying

In-storage drying methods

Low-temperature Drying

In-storage drying methods refers to those grain is dried and stored in the same container. Low-temperature drying, also known as near-ambient drying, is one of in-storage drying methods. There are four major factors which influence low temperature drying: the variability of weather, the harvest moisture content, the air flow in the storage bin and the amount of heated air. Most low-temperature dryers are built to dry grain as slowly as possible, while in the same time less spoilage on the grain. It is suggested that low temperature drying system is better operated when the average daily temperature is between 30 °C and 50 °C. Rather than control the drying air temperature, the low-temperature drying focuses on the relative humidity in order to achieve equilibrium moisture content (EMC) in all grain layers. Low-temperature drying process usually takes 5 days to several months depends on several important variables: weather, airflow, initial moisture content and amount of heat used. Among which, airflow is the key factor. Without appropriate airflow rate, spoilage will occur before drying is completed. By using heated air (LP heat, electric heat and solar heat), the relative humidity of the drying air is better controlled to achieve the desired moisture content. Usually, heated air dryer is used when the relative humidity larger than 70%. In electric heat dryers, an electrical resistant heater is usually placed before the fan to heat the airstream. In some case, a humidistat is employed to control the heater. In solar heat dryers, the drying air passes through the solar collector first to be heated (usually 10 to 12 °F rise), then enters the bin through the fan and motor. The advantages of in-storage low temperature drying are quick filling, high quality product, less equipment requirement; while the disadvantages are long drying time, electrical demand if using electric heat, high management skills and uncertain harvest moisture content.

Multiple-layer Drying

Multiple layer drying method refers to the use of LP heat or natural gas in drying corn. Compared to low-temperature methods, multiple-layer drying requires higher temperatures, which results in a shorter allowable storage time. Multiple-layer drying without stirring is the basic multiple-layer drying method, in which airstream is entered through an LP heater by a fan. Usually, the temperature rise after the LP burner is remained low in order to avoid over drying in the bottom layers in the bin. As soon as corn is dried in the bin, the burner is turned off and the fan is used to bring the corn to ambient temperature. The advantages of multiple-layer drying without stirring are little handling of corn, and bin can be used as either dryer or storage; the disadvantages are slow filling and over drying in the bottom layers (Bern and Brumm, 2010). Multiple-layer drying with stirring can not only dry grain equilibrium from top to bottom, but also decrease the air resistance of the grain. Moreover, using stirring system can avoid over drying in bottom layer problem and give a uniform grain moisture content in the whole bin. When drying is complete, the burner is turned off while the fan and stirrer are used to mix the corn to achieve equal moisture content and temperature. The advantages of adding stirring are preventing over drying and accelerating drying and allowable fill rate; the disadvantages of stirring system are additional expenses and decreasing bin capacity.

Batch Drying Methods

Bin-Batch Drying

In batch drying methods, certain amount of grain is placed first, usually 2 to 4 inches, the batch is dried and cooled later, then drying is stopped and batch is removed. The batch dryers are usually operating under this sequence and repeating this sequence for several times. The bin-batch drying methods employ a full perforated floor as the dryer. Without stirring, large variety of equipment is available and the batch can be used as both dryer and cooler, but there may be large moisture gradient from top to bottom and losing time in loading and unloading process. When adding stirring system, unequilibrium moisture content problem is avoided, however, stirrer is an added expense. When using bin-batch roof dryer, time losing problem can be solved. There is a drying floor under the bin roof and the drying fan and burner is installed high on the bin wall. When the drying process is completed, grain is put in the regular bin floor, thus unloading time is reduced. However, there is no wet grain holding in bin-batch roof dryers and there is more expense on machines.

Column Batch Drying

The column formed in this kind of dryer is made up of two vertical perforated steel sheets, which is about 12 inches thick each. The capacity of column batch dryers is too small to store grains. The advantages of column batch, stationary bed dryer are easy to move and the dyer can be used as cooler; while the disadvantages are time losing when cooling, loading and unloading and unequal moisture distribution when drying is completed. When column batch recirculating dryer is used, the moisture content variation problem is avoided, but the additional handling process may result in grain spoilage.

Continuous flow drying methods

Cross Flow Drying

Cross flow dryers is one of the most widely used continuous flow dryers. In the cross flow dryer, the airstream is perpendicular to the grain flow. Then the grain near the drying air is over dried, while on the other side, grain is under dried. Moisture gradient exists when drying is complete. In reality, the lower the airflow rate, the higher the grain moisture content variation between two sides of the column.

Concurrent Flow Drying

In the concurrent flow dryer, both the grain and air are moving in the same direction, which means the wettest grain is subjected to the hottest drying air. The kernels leave the drying region at the same temperature and the same moisture content. Energy efficiency is 40% better compared to cross flow dryer. However, the bed depth must be deeper than 12 inches depth than cross flow type. Thus, fan power requirements are high in this type of dryer.

Counter Flow Drying

In the counter flow dryer, the grain and the air are moving in opposite directions, which mean the driest grain is subjected to the hottest drying air. The kernels leave the drying region at the same temperature and the same moisture as in concurrent flow dryers. The suggesting air temperatures are less than 180 °F because the driest kernels are more likely to be damaged by hot air.

Applications of Grain Drying

Sunflower Drying

For different types of sunflowers, the preservation moisture content is different. Oilseed sunflowers are better dried to 9 percent moisture content, while bird seed sunflowers are 10 percent moisture content. Compared to corn drying, sunflowers are more easily to be dried and kept in safe storage. What's more, high temperature may not have adverse effect on sunflowers kernel, which may be the reason of the fatty acid composition. There was no evidence of damage when air was heated up to 220 °F in drying. However, fine hairs and fibers on the seed coat of sunflowers may cause fire hazard. It is suggested that remove the flaming particles first when heating the sunflowers.

Bean Drying

The seed coat of bean is quite fragile and easy to damaged in cracking and splitting, which may cause loss to the producer. Some studies on beans suggested that in order to avoid cracking, it is better to keep drying air above 40 percent relative humidity.

Corn Drying

When drying corn kernels it is important to keep in mind that cracks and fractures in the corn can lead to many problems in both the storage and processing. The major problem that occurs from high temperature drying and then rapid cooling of the grain is stress-cracking. Stress-cracking is when fractures become present in the corn endosperm. Stress-cracked kernels often absorb water too quickly, are more likely to become broken, and are increasingly susceptible to insect and mold damage during dry storage. In order to reduce the amount of grain that is lost due to stress-cracking, medium temperature and slow cooling, or natural air and low temperature drying methods should be employed.

Drying Mechanisms

In the process of drying heat is necessary to evaporate moisture from the grain and a flow of air is needed to carry away the evaporated moisture. There are two basic mechanisms involved in the drying process; the migration of moisture from the interior of an individual grain to the surface, and the evaporation of moisture from the surface to the surrounding air. The rate of drying is determined by the moisture content and the temperature of the grain and the temperature, the (relative) humidity and the velocity of the air in contact with the grain.

The moisture content falls rapidly at first but as the grain loses moisture the rate of drying slows. In general the drying rate decreases with moisture content, increases with increase in air temperature or decreases with increase in air humidity. At very low air flows increasing the velocity causes faster drying but at greater velocities the effect is minimal indicating that moisture diffusion within the grain is the controlling mechanism.

Grains are hydroscopic and will lose or gain moisture until equilibrium is reached with the surrounding air. The equilibrium moisture content (EMC) is dependent on the relative humidity and the temperature of the air;

The relationship between EMC, relative humidity and temperature for many grains has been modelled by numerous researchers; the results of which have been summarized by Brooker et al. (1974).

It is very important to appreciate the practical significance of the EMC. Under no circumstances is it possible to dry to a moisture content lower than the EMC associated with the temperature and humidity of the drying air; for example, that paddy can only dry to a moisture content of 16.7% when exposed to air at 25°C and 90% relative humidity. If paddy at a moisture content less than 16.7% is required then either the temperature of the drying air has to be increased or its humidity reduced.

The drying of grains in thin layers where each and every kernel is fully exposed to the drying air can be represented in the form:

MR = f(T, h, t); (1)

where (the moisture ratio);

MC is the moisture content of the grain at any level and at any time, % dry basis (%db);

MCe is the equilibrium moisture content (%db);

MCo is the initial moisture content of the wet grain (%db);

T is the air temperature (°C);

h is the air relative humidity; and

t is the drying time.

Empirical data have been used to determine mathematical approximations of the relationship between drying rate and air conditions. Relationships for many grains have been summarized by Brook & Foster (1981). For example, a thin layer equation for paddy (Teter 1987) is:

MR = exp.(-X * t); (2)

where X = 0.026 - 0.0045h + 0.01215T; and

Y = 0.013362 + 0.194h - 0.000177h + 0.009468T,

with h expressed as a percentage, and T in °C.

In the drying of grain in a deep bed, whilst individual kernels may all be losing moisture at different rates, the overall drying rate will remain constant for a long period. The air absorbs moisture as it moves through the bed until it becomes effectively saturated and moves through the remaining layers of grain without effecting further drying. Three zones present within a thick drying bed at an intermediate time within the drying operation. Drying takes place within a discrete zone, the size of which depends on the moisture content of the grain and the temperature, humidity and velocity of the air. Below the drying zone is the dried zone where the grain is in equilibrium with the air. Above the drying zone is the un-dried zone wherein the grain remains unchanged from its initial condition. In a shallow bed the drying zone is thicker than the bed depth and drying would occur initially throughout the bed.

The change in temperature and humidity of air as it moves through a bed of grain depends on the rate at which moisture is being evaporated from each kernel as an individually exposed element. Knowledge of the effect of grain moisture content, other grain properties, the temperature, humidity and flow rate of the air upon fully exposed kernels is essential to an understanding of how drying would proceed within a bed.

Unfortunately no theory has been developed that accurately and practically describes the thin layer drying rate. As described above many empirical relationships have been established and these have to be used in prediction of drying time. Accurate prediction of drying time is further inhibited by the variability of key factors encountered in practice, particularly so for the simple drying systems that are the most appropriate for use in developing countries. For example the moisture content of individual grains is likely to vary considerably within a batch and in the case of drying with a heater of constant heat output the temperature of the drying air will vary with changes in ambient air temperature.

Air Properties

The properties of the air flowing around the drying grain are a major factor in determining the rate of removal of moisture. The capacity of air to remove moisture is principally dependent upon its initial temperature and humidity; the greater the temperature and lower the humidity the greater the moisture removal capacity of the air.

It is important to appreciate the difference between the absolute humidity and relative humidity of air. The absolute humidity is the moisture content of the air (mass of water per unit mass of air) whereas the relative humidity is the ratio, expressed as a percentage, of the moisture content of the air at a specified temperature to the moisture content of air if it were saturated at that temperature.

The changes in air conditions when air is heated and then passed through a bed of moist grain. The heating of air from temperature TA to TB is represented by the line AB. During heating the absolute humidity remains constant at HA whereas the relative humidity falls from hA to hB. As air moves through the grain bed it absorbs moisture. Under (hypothetical) adiabatic drying sensible heat in the air is converted to latent head and the change in air conditions is represented along a line of constant enthalpy, BC. The air will have increased in both absolute humidity, Hc, and relative humidity, hc, but fallen in temperature, Tc. The absorption of moisture by the air would be the difference between the absolute humidities at C and B. (HC-HA)

If unheated air was passed through the bed the drying process would be represented along the line AD. Assuming that the air at D was at the same relative humidity, hc, as the heated air at C then the absorbed moisture would be (HD-HA), considerably less than that absorbed by the heated air (HC-HA)

Physical Properties of Grain

Comprehensive data on the numerous physical and thermal properties of grain are available in texts such as Brooker et al. (1974) and Brook & Foster (1981).

Moisture Content

Convention dictates that moisture contents of grains are usually measured on a wet basis, ie the mass of moisture per unit mass of wet grain and written as X % (wb). The alternative measure refers to the measurement on a dry basis (X%(db)) which is the mass of moisture per unit mass of completely dry grain. Conversion between the two measurements is shown in Table 1. All moisture contents given in the text are on a wet weight basis, unless otherwise stated.

Table 1: Conversion of Moisture Contents

Wet Basis %	Dry Basis %
10.0	11.0
11.0	12.3
12.0	13.6
13.0	15.0
14.0	16.3
15.0	17.6
16.0	19.0
17.0	20.5
18.0	21.9
19.0	23.5
20.0	25.0
21.0	26.5
Wet Basis %	Dry Basis %
22.0	28.2
23.0	29.9
24.0	31.6
25.0	33.3
26.0	35.1
27.0	37.0
28.0	38.9
29.0	40.8
30.0	42.8

Bulk Density

The bulk density of grain is the weight per unit volume. Moisture content has an appreciable effect on the bulk density.

Resistance to Air Flow

The energy required to force air through a bed of grain is dependent on the air flow, the grain depth and physical properties of the grain such as surface and shape factors, the kernel size distribution, moisture content, and the quantity and nature of contamination, stones, straw, weeds etc. The relation between air flow and the pressure drop generated across the bed for selected grains is illustrated. The data generally refer to clean and dry grain and correction factors of up to 1.4 are used for very wet and dirty grain (Teter 1987).

Latent Heat of Vaporization. Energy in the form of heat must be supplied to evaporate moisture from the grain. The latent heat of vaporization, Lh, for a grain depends on its moisture content and temperature and is appreciably greater than the latent heat of evaporation of water. Data for other grains have been reported by Brooker et al. (1974)

Estimation of Drying Time

A basic design procedure for the field worker is best illustrated for the design of a batch type dryer although the principles can be applied to a certain extent in the design of continuous multi-stage systems.

Assumed ambient air conditions are a dry bulb temperature of Ta and a relative humidity of RHa; from the psychrometric chart the wet bulb temperature, Twa, the enthalpy, Ha, and the absolute humidity ha can be derived. The air is heated to a selected safe drying temperature, Tb, thereby raising the enthalpy of the air to Hb.

The wet grain of equivalent bone-dry mass G has a moisture content of MCw%(db) and is to be dried to a moisture content of MCd%. A mass air flow of V is available.

The moisture, M, to be removed,

$M = G * (MCw - MCd)$; (3)

It is assumed that throughout the drying period the air will exhaust from the bed at a constant wet bulb temperature and in equilibrium with the uppermost layers of grain. Initially the exhaust air will be in equilibrium with grain at MCw moisture and finally in equilibrium with grain at MCd moisture. By superimposing equilibrium moisture content data on to the psychrometric chart for the initial and final moisture contents the humidity of the exhaust air at the beginning and end of drying can be found. An average of the initial and final exhaust air relative humidities, hea is taken for calculation of drying time, td:

An alternative, more complex but more accurate method for the estimation of drying performance is the technique based on dimensionless drying curves as initially developed by Hukill (1947). The methodology permits the estimation of the moisture content of grain at any level within the bed at any time after initiation of drying. It can be used for any grain for which EMC and thin layer drying data are available as is the case for most cereal grains.

The methodology involves the use of bulk drying curves as depicted in Figure 5.6 and calculation of three parameters, moisture ratio, time unit and depth factor.

The time unit, Y, is calculated using the equation:

where t0.5 is the half-response time, the time required for fully exposed grain to reach a moisture ratio of 0.5 under the drying air conditions employed. It can be calculated from the thin layer drying equations as in Equation 1 with MR assigned a value of 0.5.

The depth factor, D, is defined as the depth of the bed that contains the mass of grain, DM, that can be dried from the initial moisture ratio MR = 1 to a final moisture ratio MR = 0 with the sensible heat available over the period of one half-response time as the air cools to its wet bulb temperature. DM is calculated thus:

where Cp is the specific heat of air. The number of depth factors within the bed is found from the expression:

where d is the bed depth.

By transposing the drying conditions to these units and using either Figure 5.6 or Equation 8 it is possible to predict when any layer within the bed reaches a desired moisture content.

More rigorous approaches to the design of drying systems have been developed. These include the methods based on thin layer drying equations described by Brook & Foster (1981) and Brooker et al. (1974). Many of these have been developed into sophisticated simulation techniques (Brooker et al. 1974).

The drying conditions for specific grains and situations are many and varied. Drying will take place under any conditions where grain is exposed to a flow of unsaturated air. Very fast drying can be accomplished using large volumes of high temperature air but, if carried through to completion, is likely to be inefficient in energy use and liable to damage the grain by over-heating and/or over-drying. Conversely slow drying, as in sun drying in inclement weather, provides conditions for continued respiration and deterioration of the grain leading to both quantitative and qualitative losses and the growth of moulds.

Drying Efficiency

The efficiency of the drying operation is an important factor in the assessment and selection of the optimum dryer for a particular task. There are three groups of factors affecting drying efficiency:

- those related to the environment, in particular, ambient air conditions;
- those specific to the crop;
- those specific to the design and operation of the dryer.

There are several different ways of expressing the efficiency of drying, of which the sensible heat utilization efficiency (SHUE), the fuel efficiency, and the drying efficiency are the most useful.

The SHUE takes into account the sensible heat attributable to the condition of the ambient air and any heat added to the air by the fan as well as the heat supplied by combustion of the fuel. It is defined as:

The fuel efficiency is based only on the heat available from the fuel:

It can be appreciated that the fuel efficiency would be significantly different for the operation of the same dryer at two locations with widely different ambient conditions. With low temperature drying, particularly in dry climates, the heat supplied from the fuel may be less than half of the total sensible heat and the fuel efficiency may exceed 100%. Direct comparison of the performance of dryers at separate locations is not possible using the fuel efficiency.

Is the expression to be used for evaluation of dryer designs or comparison between dryers, since it is a measurement of the degree of utilization of the sensible heat in the drying air.

Foster (1973) evaluated the fuel and drying efficiencies of several types of dryers used with maize. Over a wide range of conditions, continuous-flow dryers were found to have a fuel efficiency of 38% and a drying efficiency of 51%, batch dryers 42% and 58%, dryeration 61% and 78%, and two-stage drying, 60% and 79%, respectively.

EFFECT OF DRYING ON GRAIN QUALITY

The drying operation must not be considered as merely the removal of moisture since there are many quality factors that can be adversely affected by incorrect selection of drying conditions and equipment. The desirable properties of high-quality grains include:

- Low and uniform moisture content;
- Minimal proportion of broken and damaged grains;
- Low susceptibility to subsequent breakage;
- High viability;

- Low mould counts;
- High nutritive value;
- Consumer acceptability of appearance and organoleptic properties.

Moisture Content. It is essential that the grain after drying is at a moisture content suitable for storage. As discussed the desired moisture content will depend on the type of grain, duration of storage, and the storage conditions available. It is also important that the drying operation is carried out to minimize the range of moisture levels in a batch of dried grain. Portions of under-dried grain can lead to heating and deterioration. Stress Cracking and Broken Grains. Drying with heated air or excessive exposure to sun can raise the internal kernel temperature to such a level that the endosperm cracks. The extent of stress cracking is related to the rate of drying. Rapid cooling of grain can also contribute to stress crack development.

Nutritive Value. Grain constituents such as proteins, sugars and gluten may be adversely affected when the grain attains excessive temperatures. The feeding value of grains can be lowered if inadequately dried.

Grain Viability. Seed grain requires a high proportion of individual grains with germination properties. The viability of grain is directly linked to the temperature attained by grains during drying (Kreyger 1972).

Mould Growth. Many changes in grain quality are linked to the growth of moulds and other microorganisms. The rate of development of microorganism is dependent on the grain moisture content, grain temperature, and the degree of physical damage to individual grains. Mould growth causes damage to individual grains resulting in a reduction in value. Under certain circumstances mycotoxin development can be a particular hazard.

Appearance and Organoleptic Properties. The colour and appearance as perceived by the customer and/or consumer. For example, the colour of milled rice can be adversely affected if the paddy is dried with direct heated dryers with poorly maintained or operated burners or furnaces.

Sun Drying

The traditional practice of grain drying is to spread crop on the ground, thus exposing it to the effects of sun, wind and rain. The logic of this is inescapable; the sun supplies an appreciable and inexhaustible source of heat to evaporate moisture from the grain, and the velocity of the wind to remove the evaporated moisture is, in many locations, at least the equivalent of the airflow produced in a mechanical dryer. In tropical countries, for at least several months of the year, the mean level of insolation upon the ground is more than 0.5 kW/m^2 (measured as a mean over the hours of daylight). The heat available therefore, assuming a 12 hour day, is 21.6 MJ/m^2, a quantity theoretically sufficient to evaporate 9 kg of water.

Even today, sun drying of grain remains the most common drying method in tropical developing countries. It is first employed when the crop is standing in the field prior to harvest; maize cobs may be left on the standing plant for several weeks after attaining maturity. Although not requiring labour or other inputs field drying may render the grain subject to insect infestation and mould growth, prevent the land being prepared for the next crop and is vulnerable to theft and damage from animals. Drying in the field may also be carried out after harvest with the harvested plants laid in stacks with the grain, maize cobs or panicles raised above the ground and exposed directly to the sun. Data on the drying of paddy in the field has been gathered by Angledette (1962) and Mendoza et al. (1982).

Drying on flat exposed surfaces is the most common way of drying grain after harvesting and threshing. For drying small amounts on the farm grain may be spread on any convenient area of land. Contamination with dirt cannot be easily avoided with this method and cleaner dried grain can be obtained by drying the grain on plastic sheets, preferably black.

Purpose-constructed drying floors are commonly used where there is a need to dry large quantities of grain during the season, e.g. at most rice mills. The floors are usually made of concrete or brick, these materials presenting a relatively smooth and hardwearing surface. Floors should be constructed to withstand the movement of vehicles and sloped or channelled to hasten the runoff of rainwater. The paddy is spread in a thin layer on the floors and raked at intervals, preferably 7-8 times daily, to facilitate even drying. At night the paddy is heaped into rows and covered with sheeting.

Work by Chancellor (1965) and Soetoyo & Soemardi (1979) has demonstrated that paddy can be dried from 24-26% moisture to 14% moisture at depths of 50-100 mm at a rate of 3.3 $kg/m^2.h$ for stirred paddy and 1.9 $kg/m^2.h$ for unstirred paddy. The grain can reach temperatures as high as 60°C under clear skies and the rate of drying can be extremely high. Under these circumstances kernel cracking and loss of head rice can be appreciable, particularly if paddy is dried to below 14% moisture. Covering the paddy around midday may be beneficial under particularly hot and sunny conditions. Experiments at IRRI have shown that cracking can be reduced by 25% if paddy is dried in the shade but the benefit from the improved quality is generally more than offset by the longer drying times and hence reduced throughput and increased costs.

In rainy weather, even though drying will be slow, every effort should be made to prevent wet freshly-harvested paddy from over-heating with deterioration in quality by spreading on floors rather than let it remain in heaps and sacks. Under these conditions or when there is great demand for drying space paddy can be dried to 17-18% moisture and then temporarily stored for 15-30 days before final drying.

Crib Drying

Compared with paddy, cob maize can remain at relatively high moisture contents, in excess of 20% with natural ventilation for considerably longer periods, from one to three months. The maize crib in its many forms acts as both a dryer and a storage structure. The rate and uniformity of drying are controlled by the relative humidity of the air and the ease with which air can pass through the bed of cobs. The degree of movement of air through the loaded crib is largely attributable to the width of the crib; research in West Africa has shown that crib widths should not exceed 0.6 m (Anon 1980). Guide-lines on crib design, construction and operation have been prepared by Bodholt (1985).

Solar Dryers

An improved technology in utilizing solar energy for drying grain is the use of solar dryers where the air is heated in a solar collector and then passed through beds of grain. There are two basic types of solar dryer appropriate for use with grain: natural convection dryers where the air flow is induced by thermal gradients; and forced convection dryers wherein air is forced through a solar collector and the grain bed by a fan (Brenndorfer et al. 1985).

Natural convection dryers are generally of a size appropriate for on-farm use. The dryer consists of three components, a solar collector, the drying bin and a solar chimney. For a one tonne capacity dryer the collector is 4.5 m long and 7.0 m wide with the solar absorber base of burnt rice husks or black plastic sheet covered with clear plastic sheet. The drying bin is 1.0 m long and 7.0 m wide with a base of perforated steel or bamboo matting.

The solar chimney provides a column of warm air that increases the thermal draught of air through the dryer. It is made of a bamboo frame covered with black plastic sheet. In Thailand paddy was dried from 20% moisture to 13% moisture in 1-2 days and the rice quality was appreciably greater than that from sun dried paddy. A disadvantage of the dryer is its high structural profile which poses stability problems in windy conditions, and the need to replace the plastic sheet every 1-2 years. A smaller (100 kg capacity) and simpler version of this type of dryer has also been developed (Exell & Kornsakoo 1978; Oosthuizen & Sheriff 1988). The forced convection solar dryer can be considered as a conventional mechanical drying system in which air is forced through a bed of grain but the air is heated by a flat plate solar collector rather than by more conventional means.

The performance of a flat plate collector can be quantified by calculation of the collection efficiency; the ratio of the heat gathered by the collector to the insolation incident on its surface. The collection efficiency is a function of the air velocity through the collector, the geometry of the air duct, the absorptivity of the absorption surface, and the transmissivity of the cover(s).

Considerable work has been undertaken in developing low-cost and efficient solar collectors for crop drying applications (Brenndorfer et al. 1985; Davidson 1980). The simplest type of collector is the bare plate which consists simply of an air duct the uppermost surface of which acts as the absorber plate. The covered plate collector in its many forms utilises a translucent cover above the absorber plate; Compared with the bare plate collector higher collection efficiencies are obtainable with covered plate collectors but at the expense of increased complexity and cost.

The optimum design suitable for use at farms and mills in developing countries is probably the bare plate collector which is capable of operating at a collection efficiency of 40-50% with an airflow of 0.10 kg/s.m^2. With typical insolation levels in many tropical countries of the order of 0.5 kW/m^2 such collectors are capable of providing mean daily elevations in air temperature of 5-10°C with heat outputs of 0.20-0.25 kW/m^2 of collector area. Covered plate collectors, operating at efficiencies of 60-70%, are capable of providing air temperature elevations of 10-30°C but at a lower airflow. A major advantage of the bare plate collector is that it can be easily incorporated into the roof of a dryer or storage building. Corrugated iron is a popular and inexpensive roofing material in many areas and when painted black forms an excellent solar absorber. A false ceiling can be fixed to the roof joists so forming a shallow duct running the length of the building and easily connected to a fan via ducting at one end of the building. The heat available from the collector is weather dependent and consideration should therefore be given as to whether solar energy should be the sole source for heating the air or a supplement to more conventional heating systems.

Research and performance studies on forced convection solar dryers have been reported by Bose (1978), Muthuveerappan et al. (1978) and Soponronnarit et al. (1986). Damardjati et al. (1991) have described the performance in Indonesia of a 10 tonne/day paddy dryer that incorporated a 225 m2 roof-type collector together with a moisture extraction unit (MEW, see below). Heat output from the collector averaged 60 kW over daylight hours and that from the MEU 35 kW with a mean daily elevation in air temperature of 7-9°C.

Dryeration

Originally developed for use with maize, dryeration is a combination of heated air drying and aeration cooling. In this process a tempering period is employed between a high temperature drying phase and a cooling phase. Whereas less than 1% moisture is removed if cooling is carried out immediately after drying, as much as 2% moisture can be removed if the grain is cooled slowly after tempering. Damage to the grain is reduced and drying efficiency is improved through better utilization of the residual heat in the grain for moisture removal during cooling.

Higher air temperatures can be used in the drying phase since the grain is not dried to such a low moisture content.

Two-Stage Drying

Two-stage or combination drying can be used to relieve pressure on drying facilities during peak periods. For example, paddy at moisture contents of less than 18% can be stored for up to 20 days without significant losses either in quantity or quality. In two-stage drying, grain is dried to an intermediate moisture content, 20% moisture for maize, 18% moisture for paddy, as soon as possible using any of the methods described above and then dried instore to the desired final moisture content over several days or weeks with intermittent use of ambient air or air heated by 3-5°C. Research with paddy in the Philippines (Tumambing & Bulaong 1986; Adamczak et al. 1986) has shown that, in addition to increasing throughput of the first stage dryers, there were substantial overall energy savings and no loss of quality compared to drying to 14% moisture in the conventional manner.

Pre-drying Aeration

Work in the Philippines has shown that wet paddy can be maintained in reasonable condition for 3-7 days when aerated with ambient air (Raspusas et al. 1978; de Castro et al. 1980). By aerating stacks of sacked paddy at a rate of 0.5 m3/s per tonne for eight hours a day, quality could be maintained for nine days during the dry season and two days during the wet season. Aerating in bulk with similar airflows maintained quality for 14 days and three days respectively (Raspusas et al. 1978). The length of time that paddy can remain in aerated storage without deterioration is dependent on the moisture content of the grain and ambient air conditions.

Drying of Parboiled Paddy

After parboiling, paddy contains about 35% moisture. During the parboiling process the starch is gelatinized which confers quite different drying properties to that of field paddy. It has been shown (Bhattacharya & Indudhara Swamy 1967) that in the drying of parboiled paddy, significant damage (ie kernel cracking) does not occur until the moisture content falls to 16%, regardless of the drying method or the rate of drying,. Cracking then occurs some time after the grain has cooled. The recommended drying procedure is to dry the parboiled paddy to 16-18% moisture as fast as facilities permit, temper it for four hours if warm or eight hours if cooled, and then dry in a second operation to 14% moisture. Air temperatures of 100-120°C can be used for parboiled paddy in continuous-flow dryers.

Drying of Seed Grain

If grain is destined for use as seed then it must be dried in a manner that preserves the viability of the seed. Seed embryos are killed by temperatures greater than 40-42°C and therefore low temperature drying regimes must be used. Seed grain may be dried in any type of dryer provided that it is operated at a low temperature and preferably with greater air flowrates than generally used. It is essential that batches of grain of different varieties are not mixed in any way and therefore the dryers and associated equipment used must be designed for easy cleaning. In this respect simple flat-bed dryers are more suitable than continuous-flow dryers.

Teter (1987) noted that seed paddy can be sun dried at depths of up to 30 mm but that the final stages of drying to 12% moisture should be conducted in the shade to avoid overheating and kernel cracking. Flat-bed dryers can be used with bed depths of up to 0.3 m, air temperatures not exceeding 40°C, and airflows of 1.3-1.7 m3/s per tonne of grain.

Cross-mixing between batches of different varieties can be avoided by drying in sacks in a flat-bed dryer although care must be taken in packing the loaded sacks in the dryer to ensure reasonably even distribution of airflow. Specialised tunnel dryers in which sacks or portable bins are individually placed over openings in the top of the tunnel have been developed (Teter 1987).

Air Movement

The selection and sizing of a fan to move air through a dryer is very important. The major resistance to the flow of air comes from the grain bed; the pressure drop through the bed support and ducting is of lesser effect, particularly for deep beds. The pressure drop across a grain bed is a function of the depth, the air velocity and the grain itself. It is important to note the major effect of dockage upon the pressure drop generated.

For most situations either axial-flow or centrifugal fans are used. The axial-flow fan moves air parallel to its axis and at right angles to the field of rotation of its blades. With the centrifugal fan the air enters parallel to the drive shaft, moves radially through the blades and is discharged tangentially from the housing surrounding the impeller. Axial-flow fans can be easily mounted in-line in the ducting and are relatively inexpensive but are only capable of operation against pressure drops of less than 1,500 Pa. Compared with axial-flow fans centrifugal fans can operate against higher pressure drops and are quieter in use but are more expensive.

Brooker et al. (1974) provide comprehensive information on the selection and operation of fans. It should be noted, particularly for large-scale dryers containing perhaps hundreds of tonnes of grain, that the risks of mechanical or electrical failure of the fan is likely to result in considerable losses if the fan cannot be repaired within a day or two. Consideration should be given therefore to installation of a back-up fan, particularly in locations where repair facilities are limited.

Air Heating

Heaters can be divided into two types, direct and indirect. In direct heaters the fuel is burnt in situ with the drying air so that the products of combustion pass through the drying bed with the air. Heaters of this type are less expensive and more energy efficient; however, the quality of the grain may be lowered due to contamination with combustion products, particularly if the heater is poorly maintained. In indirect heaters the combustion air does not come into contact with the drying air and a heat exchanger is used to raise the temperature of the latter. Depending on the type of heat exchanger as much as 25 % of the heat may be lost; however, there is no danger of contamination of the grain.

Air for drying can be heated by gas and oil and also solid fuels such as coal, wood and biomass residues. Oil-fired heaters are the most common for use with small on-farm dryers. Oil-fired and gas-fired heaters for all sizes of dryers are commercially available as described by Araullo et al. (1976), Brookeret et al. (1974) and Wimberly (1983). Small heaters are usually transportable and are easily positioned on the suction side of the fan so that the hot air from the heater is drawn into the plenum chamber by the fan together with ambient air.

Use of Biomass

Oil and gas are the conventional fuels employed in heated-air dryers, particularly so for small-scale operations such as the batch-in-bin dryer. The use of these fossil fuels is increasingly expensive and environmentally undesirable. The use of alternative and renewable energy sources is likely become increasingly common as new combustion technologies are developed and conventional fuels increase in cost. In many areas the residues available from grain crops, such as maize cobs and rice husks, are available in large quantities, but are generally under-utilized and present problems of disposal. Depending on the crop production systems employed other agricultural residues may be produced in the vicinity of grain drying plants and may offer alternative fuel options.

Few comprehensive measurements have been made of biomass residue availability. However, estimates have been made from the ratio of crop yield to the residue, Much of this material has current or potential use for a wide range of applications, but in many places there are underutilized resources that could be used as fuel for grain drying.

There are many different combustion systems that are currently or potentially suitable for combustion of biomass residues. Grate Furnaces. The use of grates is probably the most commonly used method world-wide. There are grate systems suitable for burning a wide range of biomass materials, including many particulate residues and straw. The grate is designed to support the biomass fuel and allow air to circulate freely through it. There are many types of this system: flat grates, both static and moving; cone grates; step grates and sloping grates (Sarwar et al., 1992). Flat grates are the simplest type and are found in the majority of log and straw burning systems. Step grates are often used to burn rice husks.

Suspension Burners. These are suitable to burn particulate agricultural residues of regular size and shape. These systems have great potential for application in developing countries. A small number of commercial and piloted systems exist (Mahin 1991; Robinson 1991).

Fluidized Bed Systems. These systems are especially suited for burning both large and small particulate agricultural residues of relatively high moisture contents. The fluidized bed furnace comprises a combustion chamber containing a sand bed acting as the heat transfer medium. Commercial units are generally large-scale and capital intensive and as such are less suited for application in developing countries.

Under-fed Stokers. These systems are also suitable for particulate biomass residues of relatively high moisture content. The biomass is transported by a screw-feeder through a specially constructed trough into the middle of the furnace. From the sides of the trough primary air is forced through the biomass mound. Secondary air is introduced near the top of the mound allowing complete fuel combustion. There has been relatively little application of under-fed stokers in developing countries.

Gasification Systems. These systems can be designed to burn wood in the form of logs and also particulate fuels. The biomass is pyrolysed to produce combustible gases and wood tars. These products of pyrolysis are then used as a fuel. Since good combustion control can be obtained with gasifiers the hot gases of combustion can be employed to direct-fired dryers. Gasifications systems are typically at the experimental or adaptive research and development stage although there has been some limited commercial success with wood and charcoal gasifiers (Breag & Chittenden 1979; Hollingdale 1983; Sarwar et al. 1992).

There are handling and combustion advantages in compressing particulate materials into a more compact form, briquettes, for use in existing furnace systems (Smith et al. 1983). Various techniques can be used for converting residues to briquettes; the piston press, the screw press, the pellet press and the manual press. The experience of briquetting agricultural residues has been mixed. Various technical problems have been encountered but the main difficulty has been the fact that, in many places, briquettes are too high in cost to compete with existing woodfuel (Eriksson & Prior 1990).

Details of the commercial availability of equipment for combustion and handling of biomass materials can be found in publications by Eriksson & Prior (1990), Sarwar et al. (1992), and the Biomass Energy Directory (Anon. 1992). A great deal of information and consideration is needed to arrive at any reasonable conclusion on the suitability of a particular combustion system for use in grain drying systems in developing countries.

5

METHODS OF GRAIN DRYING

Field Drying: Many farmers leave their crop in the field to dry. This is especially true for beans and sometimes for maize. Rice should never dry in the field. Using this method the farmer relies on the natural weather conditions (hopefully they are dry enough) to dry his or her crop.

If one is field drying maize, the stem should be broken, once the maize is ripe, so that the heads of the maize cobs can be bent towards the ground. Any rain that falls will find it difficult to enter the cob of maize if it is turned upside down.

Advantages

- Air passes freely through the field.
- Labour involved is very small.

Disadvantages

- Squirrels, rats, birds, and insects can attack the grain with ease.
- Grain can fall on the ground and spoil.
- Often the rainy season makes this type of drying too dangerous.

Sun Drying: The oldest and perhaps best known drying method is sunning the grain. The harvested grain is spread thinly and evenly directly under the sun on mats, bamboo racks, cement slabs, or tarpaulins. The grain should not be placed directly on the ground.

The grain should be spread on the ground in a layer about 5 to 7.5 cm thick (the length of your index finger). The grain should be stirred often so that wet grains are brought to the top of the layer to dry and that warm, drier grains are cooled down as they mix with the cooler grain from the bottom. The grain should always be closely watched so that when rain storms come it can be gathered in quickly.

This method dries grain quickly and cheaply. When you stir the grain often the breeze can pass in between the grain kernels to carry the moisture away. Another advantage is that the heat from the sun also bothers the insects; they will not stay in grain which is being sunned.

This method does have some disadvantages:

- The grain must be brought under shelter every night to avoid the morning dew.
- Birds, rats, and goats can enter the drying area and eat the grain.
- Drying during the rainy season is still difficult.

There is a way to improve upon the traditional sun drying method. A simple drying box can be constructed and covered with plastic sheeting.

- Construct from planks or plywood a shallow box of the following dimensions: 2 m wide by 3 m long and 20 cm high.
- Drill or cut ventilation holes in the walls of the box. The holes should be about 2 cm in diameter. Very small holes (smaller than the grain to be dried) should be made in the bottom of the box. Remember that moving air is needed to carry away moisture - the ventilation holes allow the air to pass.
- The floor of the box must be raised off the ground. Therefore, place the box on short legs about 15 to 20 cm high.

Now a drying roof must be constructed.

- Buy a 3 by 5 m sheet of plastic in the market.
- Construct a simple wooden frame with dimensions of 2 m by 3 m from thin sticks which will fit on top of the 2 m by 3 m drying box.
- Stretch the plastic sheeting over the roof frame and attach it.

Now the box can be placed in the sun. Then the box is filled with maize 5 to 7.5 cm deep. The plastic roof is then placed over the box and tied to it so that the wind will not blow it away. The inside of the box will become very hot air can pass over and through the grain by the ventilating holes in the side walls and floor. The grain will dry very fast. Even if rain falls the roof will keep the grain out of the rain. Remember to stir the maize from time to time. Remove the grain each evening so the grain will not sweat in the box.

This 2 m by 3 m box with 6 to 7.5 cm of grain spread on the floor will hold between 225 and 300 kg of dried maize. A farmer can build any size of box he wishes.

(b) Natural Air Drying

Many farmers dry their maize using many different types of natural air drying methods. Each of these methods must let the air pass around the kernels to dry them. Therefore, to be completely successful the grains must be exposed to the air.

When drying maize this means that the husks, or leaves, covering the cob of maize should be removed. These leaves are bad for drying for two reasons:

- Leaves keep the air from directly touching the grain kernels. The moisture in the kernels is trapped inside and is more difficult to remove through the leaves.

- The leaves themselves also contain moisture which makes drying take longer and can spoil the grains inside.

However, many farmers like to keep the leaves on the maize cob because they feel that it protects the maize from insects. This is somewhat true. but many insects can still enter through the leaves and attack the grains.

It is impossible to. tell if insects are in the maize if the leaves are covering the cob. Also, if you wish to use an insecticide the leaves will stop the chemical from reaching the insects in the grain.

If you are drying the grain in the kitchen and the smoky flavour enters the maize then some leaves can be left to keep the smoke off the grain kernels.

If you do not remove the leaves from your maize during drying, then you can try an experiment of your own. Select 30 or 40 maize cobs and remove the leaves; leave the rest of your maize as before. Now dry the crop in your normal way. Check the maize often so that you can see which ones dry faster. Normally, the maize with the leaves removed dries faster and better then maize with the leaves covering the cob.

Eave Drying: With this method the maize is hung in bunches under the eave of the roof to dry. A few leaves are left on the maize cob, these leaves are used to tie the maize in a bunch. Then the bunches are hung from a pole under the eaves. Air can pass freely around the maize cobs. The eave keeps all rain water off the grains. Drying normally takes place over a period of two months in the highland areas.

The disadvantage of this method is that rats and birds can sometimes worry the grain. Also sometimes the farmer does not have space to dry his entire crop under his eaves.

Maize Crib Drying: The maize crib can only be used for the dry season crop. However care must be taken to keep falling rain off the maize inside the crib. In the rainforest zone the crib cannot be used during the rainy season.

Remember that the crib should be 60 cm (2 feet) wide so that air passes through it. The floor should be one meter off the ground to keep rats from jumping up into the crib. Do not build it under a tree, rats can climb the tree and drop into the maize. Rat guards should be placed on all legs. The long sides of the crib should face the sun (east and west) so that the sun's heat is also used to dry the grains.

Birds and insects are hard to control with this drying method. Also, drying during the rainy season is very difficult.

Construction of a Maize Crib

Step 1: Dig the ground holes. Set the wall and floor supports

Step 2: Attach the wall braces and put the floor boards in place

Step 3: Tie the bamboo sticks to the wall supports. Attach a roof of zinc or thatch.

(c) Direct Fire Drying

Kitchen Drying: Another effective way to dry grains is by placing the grain in the attic of the kitchen above the cooking fire.

In this case a few leaves can be left on the maize cob so that the smoke does not give the grain a bad taste, although some people like the smoky flavour. The heat and smoke can also keep some of the insects away. This method was developed during the pre-colonial times when all houses had grass roofs. The heat and smoke and moisture from the i grain easily passed out through the grass roof. However, with so many houses using zinc roofs today there is a problem. The heat and moisture are trapped under the zinc roof.

Drying in the kitchen

Zinc roofs trap heats and moisture; grass roofs allow it to escape

A few other problems with this method are as follows:

- As soon as the fire is put out and the smoke stops insects can fly back to the kitchen - they will leave only after the fire is started again. Space is limited under the roof - the capacity is often too small to dry the entire crop.
- Fire wood is very expensive.

Oil Barrel Method: A very fast drying method is with the oil barrel dryer. These are constructed exactly like Cocoa Dryers. However, the cost of the dryer, the cost of the wood to dry the grain, and the labour and time involved make this method too expansive for most farmers. Also the heat is often too high for the drying of rice, beans, and seed grain.

Grain Dryers: Selection, Design, Specifications and Testing

Batch-in-Bin Dryers

The small capacity version of the batch-in-bin dryer, otherwise known as the flat-bed dryer, has been developed for farm- or village-level use. Its capacity is of the order of 1-3 tonnes/day with drying times of 6-12 hours.

The flat-bed dryer is simple to construct using easily available and inexpensive materials and easy to operate with unskilled labour. The walls of the drying bin can be constructed of wood, brick or metal. The floor of the drying chamber is preferably made from fine wire mesh, suitably supported, or perforated metal. If these are not available then sacking spread over a coarser but stronger wire mesh can be used. To facilitate an even airflow through the bed the length of the drying chamber should be 2-3 times the width. The height of the plenum chamber is of the order of 0.3 m. Unloading ports can be fitted at intervals in the walls of the drying chamber.

In order to prevent excessive moisture gradients through the bed, the depth of grain in the bin is relatively shallow, 0.4-0.7 m. and the air velocity is usually of the order of 0.08-0.15 m/s for maize and 0.15-0.25 m/s for paddy. The temperature of the air is selected according to the desired safe storage moisture content of the grain. For the drying of paddy in tropical areas an air temperature of 40-45°C is usually used, with a heater capable of raising the air temperature by 10-15°C. With such bed depths and air velocities the pressure drop over the bed is relatively low, 250-500 Pa, and therefore simple and inexpensive axial-flow fans can be used. Typically power requirements are 1.5-2.5 kW per tonne of grain for a belt-driven fan powered by a petrol or diesel engine.

Operation of flat-bed dryers invariably results in a moisture gradient between the lower layers and the higher layers of the bed (Soemangat et al. 1973). This problem can be reduced by careful selection of drying temperature and airflow conditions but, even so, gradients of 34% moisture are to be expected. Turning of the grain in flat-bed dryers at intervals can alleviate the problem but this extends the drying time and requires additional labour.

The flat-bed dryer is easily loaded from sacks by hand. However, unloading the dried grain into sacks can be time- and labour-consuming; placement of the drying bin on a tilting frame has been investigated (Wimberly 1983) but this incurs additional costs.

Dryers of this type have been developed in many countries and designs are available from the University of the Philippines at Los Baños (UPLB), Los Baños, Philippines and the International Rice Research Institute (IRRI), Manila, Philippines.

IRRI have also developed a vertical batch-in-bin dryer which operates more efficiently than the flat-bed dryer. It differs from the latter in that the airflows horizontally through the bed on either side of the plenum chamber and exhausts through slatted sides. The bin is easily unloaded by removing the slats. Details are available from IRRI.

Both direct and indirect heaters can be used with the flat-bed dryer. Solar air heating can also be an option. The waste heat from the engine used to power the fan can be used (Esmay & Hall 1973; Soemangat e' al. 1973; Teter 1987). Heating of the air by 5-10°C using waste engine heat is possible but the engine exhaust gases should not be drawn through the grain; the exhaust gases should be ducted outside the housing around the engine. A development of this principle is the moisture extraction unit (MEW) in which the fan is directly driven by the engine and the air is drawn over and around the engine block and exhaust pipe.

Large capacity batch-in-bin (or in-store) dryers can be used in cooler dryer areas. The advantage of this technique is that the bin is used for both drying and storage with savings in both capital and operating costs. With heated air at a temperature of 40-45 °C, bed depths of 2-3 m can be used with air velocities through the bed not exceeding 0.08 m/s. Since drying times to achieve reductions in moisture content of 5-10% can be of the order of 20-40 days, this method should not be used in humid areas with grain of moisture contents greater than 18% because of the risks of sprouting and mould growth in the upper layers of the bed.

Large batch-in-bin dryers are usually round or rectangular and range in capacity from ten to several hundred tonnes. With large bins, air distribution ducts at the base of the bin are used rather than a plenum chamber. The ducts can be semi-circular, rectangular or triangular. To ensure good air distribution through the bed, ducts should be spaced from each other at a distance of half the depth of grain and one quarter the depth from the end and side walls. Air velocity through the ducts should not exceed 5 m/s because of pressure drop factors. More than one fan can be used to provide the airflow required. Detailed information on duct design and airflow distribution is presented by Brooker et al. (1974).

In-bin layered drying with ambient air can be performed with confidence in locations where the relative humidity of the air is less than about 70%. An initial layer of grain, 0.6-0.9 m deep, is loaded into a storage bin, 5-10 m deep, and further layers are added as drying proceeds. Over-drying of the grain is minimized because of the low air temperature. In the USA an airflow of 0.025-0.06 m3/s per tonne was used to dry paddy from 20% moisture to 16% moisture within 14 days when ambient temperature ranged from 18-24°C (Houston 1972). Careful and skilled management is required to ensure that each layer is dried before the succeeding layer is loaded into the bin.

Some success has been reported in Indonesia for the drying of paddy from 18% moisture to 13% moisture (Gracey 1978; Renwick & Zubaidy 1983). The latter also demonstrated that field-wet paddy (24% moisture) could be dried safely in bulk to 18 % moisture by continuous aeration (24 in/day) with ambient air, regardless of the humidity of the air. Subsequent drying to 14% moisture or less was accomplished by drying at times of lesser humidity and with the addition of waste engine heat.

There can be a need to dry grain in sacks in certain instances: for example, at central drying facilities where farmers wish to retain access to their own grain. Stacks of sacks are laid over air distribution ducts with no need for a conventional drying bin. Drying proceeds in much the same manner as for bin drying. After use, the air distribution ducts can be dismantled easily to allow use of the building for other purposes.

Re-circulating Batch Dryers

This type of dryer avoids the problems of moisture gradients experienced with bin dryers by re-circulating the grain during drying. The dryer is a self-contained unit with an annular drying chamber, 500 mm thick, around a central plenum chamber, a fan and heater, and a central auger for transporting the grain from the bottom to the top. When drying is complete the grain is discharged from the top. Most dryers of this type are portable and can be moved relatively easily from farm to farm.

Air temperatures of 60-80°C are employed with air flowrates of 0.9-1.6 m3/s per tonne of grain, twice that used in flat-bed dryers (Wimberly 1983). However, since the grain is only exposed to the flow of hot air for relatively short times within each cycle, too rapid drying rates are avoided and moisture distribution within individual grains is equalised during the period the grain remains in the non-drying sections at the top and bottom of the dryer. Control of the drying rate can be effected by adjusting the auger speed to regulate the flow of grain through the dryer.

Another version of a re-circulating batch dryer is rectangular with drying chambers on either side of the heater, fan and plenum chamber. Under each drying chamber are horizontal screw conveyors that collect the grain and return it to a screw auger at one end that lifts the grain to a holding section at the top. A screw conveyor in the holding section distributes the grain evenly along each drying chamber.

The capital cost of re-circulating batch dryers is considerably greater than batch-in-bin dryers because of their greater complexity and incorporation of handling and conveying devices. However, throughput is greater due to the shorter drying times and the quality of the dried grain is likely to be higher. Re-circulating batch dryers require specialist skills for construction and trained operators for successful operation and therefore are not generally suitable for operation by small-scale farmers or enterprises.

Continuous-flow Dryers

Continuous-flow dryers can be considered as an extension of re-circulating batch dryers. However, rather than the grain re-circulating from bottom to top, as in the latter, the grain is removed from the bottom, in some systems, cooled, and then conveyed to tempering or storage bins. In their simplest form continuous-flow dryers have a garner (or holding) bin on top of a tall drying compartment. With some dryers a cooling section is employed below the drying compartment in which ambient air is blown through the grain. At the bottom of the dryer is the flow control section that regulates both the circulation of grain through the dryer and its discharge.

There are three categories of continuous-flow dryers based on the way in which grain is exposed to the drying air:

- crossflow, in which the grain moves downward in a column between two perforated metal sheets while the air is forced through the grain horizontally. Dryers of this type are relatively simple and inexpensive, but, unless mixing systems are incorporated, moisture gradients are set up across the bed;
- counterflow, which employs a round bin with an unloading system at the base and an upward air flow. These dryers are relatively efficient since the air exhausts through the wettest grain. Bed depths of up to 3-4 m can be used;
- concurrent flow, which is the reverse of counterflow drying in that the air moves down through the bed. High air temperatures can be used since the air first comes into contact with wet, and sometimes cold, grain. Drying is rapid in the upper layers but slower at the bottom with some tempering action. Bed depths of at least a metre are used;

Probably the most commonly used continuous-flow dryer is the crossflow columnar dryer, which can be classified as non-mixing and mixing types.

In one version of a non-mixing dryer (Figure 5.14. Continuous Flow Dryers.), drying takes place between two parallel screens, 150-250 mm apart on either side of the plenum chamber. The air escapes from the dryer through louvres on either side of the dryer. The flow rate of grain through the dryer is controlled by a regulator gate at the base of the drying column. Since the grain flows plug-like through the drying section the layer of grain closer to the plenum chamber is dried by hotter and drier air than is the grain on the outside. However, mixing is effected to a fair degree when the grain is discharged and conveyed to tempering and storage bins. Air temperatures of 45-55°C and airflows of 2-4 m3/s per tonne of grain are used. Flow problems can be encountered with very wet and dirty grain as the grain may clog. Teter (1987) notes that if very wet paddy is to be dried then the grain should be cleaned and also pre-dried to at least 22% moisture before a non-mixing dryer can be used.

In one design of the mixing type of continuous-flow dryer, a baffle system facilitates the mixing of grain and avoids the development of moisture gradients across the drying bed. Higher air temperatures, 60-70°C, can therefore be used without damaging the grain. Unless screens are fitted on the outside of the drying section lower airflows, 1-1.5 m3/s per tonne of grain, have to be used to avoid grain being blown out of the dryer.

Another design of this type is the LSU (Louisiana State University) dryer. In this version the drying section consists of a vertical compartment across which rows of air channels are installed. One end of each channel is open and the other closed. Alternative rows are open to the plenum chamber and intervening rows to the exhaust section. Alternate rows are also offset such that the channel tops divide the moving stream of grain as it descends providing considerable mixing.

Further information on continuous-flow dryers has been presented by Bakker-Arkema et al. (1982), Fontana et al. (1982) and Houston (1972). As can be appreciated many of these dryers are large and complex structures and are usually designed and constructed by specialist firms.

Compared with batch-in-bin dryers and re-circulating batch dryers, continuous-flow dryers offer the largest drying capacity. When large volumes of wet grain are to be dried in a single site these are the types to be considered first. They are most commonly used in a multi-pass drying operation. Investment costs are high but because of the large throughputs operating costs per tonne can be lower than the larger batch-in-bin dryers and re-circulating dryers.

In a multi-pass drying system, continuous-flow dryers are used in association with tempering bins. During each pass through the dryer the grain is dried for 15-30 minutes with a reduction in moisture content of 1-3%. Drying at this rate sets up moisture gradients within the individual grains. After each pass the grain is held in a tempering bin where the moisture within the kernel equalises as moisture diffuses from the interior of each kernel to the surface. The combination of rapid drying and tempering is repeated until the desired moisture content is attained. Using this procedure the actual residence time of the grain within the continuous-flow dryer is of the order of 2-3 hours to effect a 10% reduction in moisture. Selection of the number of passes is a compromise between the dryer efficiency, ie fewer passes, and grain quality, ie longer drying time. Tempering periods are usually 424 hours in duration. The tempering bins may be aerated with ambient air to cool the grain with some slight moisture removal.

It is vital that the operation of drying with tempering is carefully planned and managed to ensure maximum throughput and efficiency. This usually means that the plant is operated 24 hours a day with two or more batches of grain being dried at a time. Well trained management and staff are essential.

Fluid Bed Drying

This type of dryer in which individual grains are suspended and sometimes transported by air moving at high velocity, 2-3 m/s, can produce very evenly dried grain. Recent research in the Philippines (Sutherland & Ghaly 1990; Tumambing & Driscoll 1991) has indicated that the fluid bed dryer has promising potential for the rapid first-stage drying of paddy to 18% moisture in two-stage drying. Paddy at a bed depth of 100 mm can be dried from 24% to 18% moisture in 15 minutes with air at 100°C and a velocity of 2 m/s, with no adverse effects on quality. However, due to the high air velocities required to fluidise the paddy, power requirements for the fan are high and the thermal efficiency is low compared to conventional (fixed bed) drying. Re-cycling of the exhaust air was identified as a potential means to improve the thermal efficiency.

Conduction Drying

Work at IRRI (Stickney et al. 1983) has investigated the use of a heated floor dryer. This consisted of a metal floor heated to 50-90°C by circulation of water heated by a furnace burning agricultural wastes. Paddy at depths of up to 60 mm could be dried from 22-26% moisture to 18% moisture in 1-2 hours depending on the floor temperature. Frequent raking was necessary but no parboiling effects were recorded and grain breakage was generally lower than that of sun dried paddy. This method is also considered as an option for the first (rapid-drying) stage of two-stage drying.

The Warehouse Dryer

This dryer has been developed (Jeon et al. 1984) for use with a wide range of crops including maize and paddy. Its particular feature is the use of a wind-powered vortex flow inducer as an alternative to conventionally powered fans for generating increased airflow over and around the drying grain. The flow inducer is mounted centrally on the roof of the dryer building and draws air, heated by a furnace and heat exchanger, through the drying bins or trays positioned in the middle of the dryer. The performance is governed by the velocity of the prevailing wind.

Rotary Drying

This method of drying has been researched at the IRRI in Philippines and also at the AIT in Thailand. Small dryers for farm use were developed at the IRRI as reported by Espanto et al. (1985). A directly-heated version consists of a perforated iron drum (0.6 m in diameter and 0.9 m long) mounted over a portable stove. The interior of the drum is fitted with flights to facilitate mixing and uniform heat transfer. The drum is rotated manually.

An indirect-heated dryer is constructed from a 2001 oil drum also mounted over a stove. Air is passed through the drum by a fan.

The performance of both dryers was very similar. Batches of 25 kg of paddy at 28% moisture can be dried to 18% moisture in 50-60 minutes and batches of maize from 33% moisture to 18% moisture in 80 minutes. Larger versions of the indirectly- heated dryer have been built (Jeon et al. 1990). With paddy, milling quality was improved relative to sun drying but the viability of the grain was greatly reduced due to the high temperatures attained at the drum surface. Similar results were obtained with the AIT dryer as reported by Jindal & Obaldo (1986).

Microwave and Infrared Dryers

When grain is irradiated by electromagnetic energy high temperature potentials are generated between the interior and surface of individual grains. Moisture therefore migrates to the surface where it evaporates to the surrounding air. The rate of airflow necessary is that required to absorb the moisture and not as the provider of latent heat. This reduction in airflow would minimize the dust and other pollutants discharged to atmosphere. More uniform drying is possible compared with conventional heated-air drying. However the capital cost and energy consumption of the microwave equipment necessary is considerable. Radajewski et al. (1988), in Australia investigated, using simulation techniques, the use of microwave heating as a means of pre-heating wheat before drying and concluded that the reduction in drying time could not offset the power consumption required for microwave heating. Infrared heating systems are similarly expensive, and since infrared radiation only penetrates superficially it is necessary to agitate the grain thereby exposing all the surface area to the radiation, thereby incurring additional cost.

6

IRRIGATION TECHNIQUES

Land conversion involves the use of previously undeveloped land to accommodate development. Infrastructure includes the capital improvements necessitated by growth encompassing both roads and water/sewer facilities. Real estate development costs typically are considered on a cost-per-unit basis for a variety of types of residential and nonresidential units (e.g., single-family detached and attached homes, retail and office space). In terms of public service fiscal effects, development is directed toward areas of excess service capacity, as opposed to locations that would have to expand their public services and infrastructure.

The research summarized here contrasted 2 alternative development futures for the United States. One alternative is current development trends-labeled conventional development, or sprawl-extended into the future. Development of this type typically includes subdivision-style residential development and strip nonresidential development. The focus is on noncontiguous land development, including residential development in the form of 0.33- to 1.0-acre lots and nonresidential strip development involving floor-to-area ratios of 0.20 or less.3 Such land development patterns would continue earlier trends of consumption of agricultural and sensitive environmental land, significant road/pavement construction, and high amounts of water and sewer infrastructure provision. Conventional development has been reported as contributing to both high real estate development costs for new development and negative fiscal effects on host public service jurisdictions.

However, sprawl development also has benefits. For example, people have access to less expensive, single-family homes on large lots situated away from urban centers and a greater opportunity for participation in governance owing to the high number of small jurisdictions found in these peripheral areas. In addition, public service costs are lower in such areas as a result of the reduced need for a deep public service base.

In the analyses described here these benefits were taken into account (i.e., both single-family purchase costs and public service expenditures are lower in areas away from urban centers), although increased participation in governance was not measured. However, managed growth remains more costefficient in the various cost comparisons undertaken. This observation is true in the first case because a greater array of housing is offered in areas close to urban centers, resulting in overall housing costs being lower in these areas in spite of lower peripheral single-family housing costs. The observation is true in the second case because tax rates in urban center areas are higher, generating even more revenues than are compensated for through reduced expenditures on a more limited array of services in areas farther out. In other words, the analyses described here encompassed and even validated the supposed benefits of sprawl, but managed growth, in spite of sprawl's benefits, remains more advantageous.

The second alternative is a more compact form of growth. Managed or "smart" growth seeks to contain most new growth around existing urban centers and to limit development in peripheral rural and sensitive environmental areas. It further seeks to reduce road construction and water/sewer infrastructure provision through higher-density development and, in some cases, mixed-use development. This goal is accomplished through increasing the share and density of development in areas closer to existing development and decreasing the share and density of development in the outer, more rural and undeveloped sections of a metropolitan area.

The research discussed here also involved a recognition that there are disadvantages associated with managed growth in the resource areas under investigation, including (1) increased housing costs owing to the land development limitations posed by managed growth, (2) extra governmental costs stemming from the administrative requirements of imposing a growth management regimen, and (3) the thwarting or driving away of development potential because of an overcontrolled real estate market. Again, these disadvantages were taken into account in the analyses.

All future development was accommodated within the assumptions of the present analysis-no development was denied-and real estate development costs were found to be lower under managed growth. Thus, even after price increases in areas where development would be limited were accounted for, overall real estate development prices would be lower. Furthermore, encouraging development in areas where government is established decreases the costs of servicing this development. The excess capacity of public services found in these locations allows public services to be extended without the hiring of new personnel.

Finally, development incentives in the form of increased density were included in the managed growth scenario. These incentives provide an impetus for developers to build in areas closer to existing development. Reduced competition between developers building farther out ensures that many of those buying houses will continue to go there. These 2 benefits support the claims of equal development in a metropolitan area under each scenario. Development is not driven away.

Before examining the benefits of managed growth versus conventional development, it is necessary to ask, "Why is conventional development, with its resource consumption excesses, so popular?" The answer involves 3 considerations: market, policy, and personal. From a market perspective, as long as society is willing to provide adequate public services at a great distance from where these services currently exist at less than full cost to the recipient, single-family housing on large lots will be less expensive at greater distances from the center of a metropolitan area. To the degree that this development form continues to be desired, people will purchase in these locations. As this question relates to policy, if mortgage interest costs and real estate taxes continue to be tax deductible, and gasoline prices are maintained at low levels because the provision and upkeep of roads are not adequately represented in this tax, there will be a continued quest for unrestrained development in the United States.

Finally, from a personal perspective, if public safety is enhanced in peripheral locations and property taxes are lower because the social safety net need not be cast as wide, again the quest for peripheral development will continue. Market, policy, and personal choices support conventional development or sprawl because resources are relatively plentiful and no one is advocating for society's needs. Individual maximization is not societal maximization, and in the short run individual maximization involves bearable negative societal costs. However, these societal costs will increase as resources become scarce, and a different public mood will result.

Models Used to Measure Resource Differences

The present analysis sought to determine 25-year (2000-2025) growth projections under the alternatives of conventional development and managed growth. The models used in the analysis create growth simulations and their effects on the basis of empirical relationships. Most of these are simple mathematical relationships that have been assembled through knowledge about how growth affects required resource supplies. These relationships are explained in more detail within each resource category.

The models begin with 2 identical projections of population and employment levels for a given US economic area made by Woods & Poole Economics Inc, a commercial vendor whose projection accuracy rates have been shown to be equivalent to those of the US census. Under alternative scenarios, different growth projections are made for different counties (the geographic unit of analysis in this study): rural and undeveloped counties are expected to experience more growth under conventional development, and urban and suburban counties are expected to experience more growth under managed growth. These projections are subjected to a housing model that changes them to different types of residential and nonresidential space and ultimately to land conversion, infrastructure, real estate development, and fiscal impact models that project land, road, and water/sewer infrastructure demands and housing/public service costs.

The resource requirements for nationwide growth over 25 years are large; thus, differences in resource consumption caused by alternative growth patterns are also large. Numbers are presented as they appear in terms of their mathematical extensions. As a result, they are both large and detailed. The accuracy rate that surrounds these numbers is generally about 10%. Projections of effects are not rounded; one can verify the mathematical extensions produced by the impact of each development scenario.

The County as the Geographic Unit of Analysis

Although the county is not a perfectly satisfactory geographic unit owing to size variations and the impossibility of controlling for intracounty development differences, the county used in this study to identify and analyze locations of residential and nonresidential development. The county was selected because (1) it is the smallest consistent noneducational unit of US government for which a large number of the demographic variables required for this analysis are consistently available; (2) a reasonable number of subdivisions of the United States as a whole are represented within this designation (approximately 3100 individual subjurisdictions); (3) projections of major demographic variables are available at this level; and (4) the problems of rapid growth and the economic and political incentives needed to redirect this growth increasingly require responses from governments of geographic areas that are larger than municipalities yet more accountable than regions with limited governance powers.

The limitations of using the county as a basic unit of analysis relate to size. We compensated for size differences by using county subunits (urbanized areas) to divide counties into smaller and different parts. Thus, even though the county is the basic unit of analysis, in about 1000 of the 3100 counties, subunits in the form of urbanized areas exist and are used as a centralized development area to redirect within-county growth. Intra county growth is targeted toward urbanized areas and away from rural areas in the same way that intercounty growth is targeted to suburban and urban counties and away from rural and undeveloped counties. Thus, although the county is the focus of most projections, these projections actually incorporate subcounty areas in instances in which these areas were found to exist.

Any plausible redirection in growth from 1 area to another must account for existing economic interrelationships among counties. In other words, if households and jobs are to be directed elsewhere to control sprawl, those locations must fall within the commuting patterns that link households and jobs in a given area. The economic area aggregation developed by the Bureau of Economic Analysis fulfills this requirement. The economic area is one of the few data aggregations that merges metropolitan and nonmetropolitan locations into an economically related geographic area. It can be viewed as similar to an extended metropolitan area.

Within a given economic area, counties are classified according to their existing levels of development. The land use development classifications are urban center, urban, suburban, rural center, rural, and undeveloped. These classifications, which are based on gross population density (population divided by surface area), identify a county as more or less developed relative to the other counties of its region and are an important consideration in determining where populations are to be redirected under the controlled growth scenario. The gross density thresholds for the classifications vary according to state gross density groupings, which in turn differ by region of the country; for example, an urban location in New Jersey would have a gross density 10 times that of an urban location in Wyoming.

Defining Sprawl Generically and Empirically in US Counties

One of the most difficult tasks of this research was, in fact, to identify and characterize the development of sprawl. These difficulties are compounded if an attempt is made to define sprawl empirically. Sprawl is lowdensity, leapfrog development characterized by strip form along major thoroughfares and tributaries at the periphery of a metropolitan area. In other words, sprawl is significant residential or nonresidential development in a relatively pristine setting. In nearly every instance, this development is low density, it has leapt over other development to become established in an outlying area, and its very location indicates that it is unbounded. This definition of sprawl is well documented in the planning literature. The definition of sprawl used in the present analysis was based on rapid and significant residential or nonresidential growth, or both, in rural and undeveloped counties. Numerous analyses of statistical population growth rates have defined the upper growth quartile as an indicator of significant growth or at least as separating significant and average growth. Here the upper quartile within an economic area was used to define significant growth.

All analyses of growth must contain both a relative and an absolute dimension. If absolute growth is taking place in a rural or undeveloped location at a multiple of the level of normal growth (in the present case, 160% of the national average absolute level), the location is defined as "sprawling," regardless of its relative rate of growth within the overall economic area. Therefore, sprawl is taking place in nonurban locations (rural and undeveloped counties)1 if either of the following sets of criteria are met: (1) the county's growth rate is in the upper quartile of the economic area's annual county household and employment growth rates, the county's growth rate exceeds the national average annual absolute growth rate, and the county's absolute level of growth exceeds 40% of the national average annual absolute growth, or (2) the county's absolute level of growth exceeds 160% of the national average annual absolute growth. Given this definition of when and where sprawl takes place, sprawl will occur in 742 of 3091 counties during the 25-year projection period (2000-2025), or 24% of all counties in the United States. In addition, although sprawl development may be present in a significant sense in only 24% of counties, it will affect about 13.1 million of the 23.5 million new households projected for the period 2000 to 2025.

The Alternative Growth Scenario: Managed Growth

Managed growth is defined as limiting a significant share of development to already-developed counties or to areas as close to already-developed locations as possible. This limiting process takes place in 2 ways. The first method limits the amount of growth taking place in the outer counties of an economic area by redirecting it to inner counties. This is accomplished by establishing an urban growth boundary around the developed (urban and suburban) counties and allowing only a portion of the growth to occur in the less developed (rural and undeveloped) counties (intercounty sprawl control).

A second method of limiting sprawl curtails the outward movement of growth in a single county (intracounty sprawl control). This is accomplished by establishing an urban service area in a county and containing most of the growth within that area. A boundary is established around the existing concentration of growth (the urbanized area), and the remainder of the county is "protected" from significant development owing to the unavailability of adequate public services. These 2 methods of limiting sprawl combine to form the managed growth scenario. Over the 25-year projection period 2000 to 2025, the United States will grow under either scenario by 23.5 million households and 49 million jobs.

Irrigation may be defined as the science of artificial application of water to the land or soil. It is used to assist in the growing of agricultural crops, maintenance of landscapes, and revegetation of disturbed soils in dry areas and during periods of inadequate rainfall. Additionally, irrigation also has a few other uses in crop production, which include protecting plants against frost, suppressing weed growing in grain fields and helping in preventing soil consolidation. In contrast, agriculture that relies only on direct rainfall is referred to as rain-fed or dryland farming. Irrigation systems are also used for dust suppression, disposal of sewage, and in mining. Irrigation is often studied together with drainage, which is the natural or artificial removal of surface and sub-surface water from a given area.

Irrigation is also a term used in medical/dental fields to refer to flushing and washing out anything with water or another liquid.

History

Archaeological investigation has identified evidence of irrigation where the natural rainfall was insufficient to support crops.

Perennial irrigation was practised in the Mesopotamian plain wherby crops were regularly watered throughout the growing season by coaxing water through a matrix of small channels formed in the field.

Ancient Egyptians practised Basin irrigation using theflooding of the Nile to inundate land plots which had been surrounded by dykes. The flood water was held until the fertile sediment had settled before the surplus was returned to the watercourse. There is evidence of the ancient Egyptian pharaoh Amenemhet III in the twelfth dynasty(about 1800 BCE) using the natural lake of the Faiyum Oasis as a reservoir to store surpluses of water for use during the dry.

Terrace irrigation is evidenced in pre-Columbian America, early Syria India and China. In the Zana Valley of the Andes Mountains in Peru, archaeologists found remains of three irrigation canals radiocarbon dated from the 4th millennium BCE, the 3rd millennium BCE and the 9th century CE. These canals are the earliest record of irrigation in the New World. Traces of a canal possibly dating from the 5th millennium BCE were found under the 4th millennium canal. Sophisticated irrigation and storage systems were developed by the Indus Valley Civilization in present-day Pakistan and North India, including the reservoirs at Girnar in 3000 BCE and an early canal irrigation system from circa 2600 BCE. Large scale agriculture was practiced and an extensive network of canals was used for the purpose of irrigation.

Ancient Persia (modern day Iran) as far back as the 6th millennium BCE, where barley was grown in areas where the natural rainfall was insufficient to support such a crop. The Qanats, developed in ancient Persia in about 800 BCE, are among the oldest known irrigation methods still in use today. They are now found in Asia, the Middle East and North Africa. The system comprises a network of vertical wells and gently sloping tunnels driven into the sides of cliffs and steep hills to tap groundwater. The noria, a water wheel with clay pots around the rim powered by the flow of the stream (or by animals where the water source was still), was first brought into use at about this time, by Roman settlers in North Africa. By 150 BCE the pots were fitted with valves to allow smoother filling as they were forced into the water.

The irrigation works of ancient Sri Lanka, the earliest dating from about 300 BCE, in the reign of King Pandukabhaya and under continuous development for the next thousand years, were one of the most complex irrigation systems of the ancient world. In addition to underground canals, the Sinhalese were the first to build completely artificial reservoirs to store water. Due to their engineering superiority in this sector, they were often called 'masters of irrigation'. Most of these irrigation systems still exist undamaged up to now, in Anuradhapura and Polonnaruwa, because of the advanced and precise engineering. The system was extensively restored and further extended during the reign of KingParakrama Bahu (1153-1186 CE).

The oldest known hydraulic engineers of China were Sunshu Ao (6th century BCE) of the Spring and Autumn Period and Ximen Bao (5th century BCE) of the Warring States period, both of whom worked on large irrigation projects. In the Szechwan region belonging to the State of Qin of ancient China, theDujiangyan Irrigation System was built in 256 BCE to irrigate an enormous area of farmland that today still supplies water. By the 2nd century AD, during the Han Dynasty, the Chinese also used chain pumps that lifted water from lower elevation to higher elevation. These were powered by manual foot pedal, hydraulic waterwheels, or rotating mechanical wheels pulled by oxen. The water was used for public works of providing water for urban residential quarters and palace gardens, but mostly for irrigation of farmland canals and channels in the fields.

In 15th century Korea, the world's first water gauge, uryanggye , was invented in 1441. The inventor was Jang Yeong-sil, a Korean engineer of the Joseon Dynasty, under the active direction of the king, Sejong the Great. It was installed in irrigation tanks as part of a nationwide system to measure and collect rainfall for agricultural applications. With this instrument, planners and farmers could make better use of the information gathered in the survey.

Types

Various types of irrigation techniques differ in how the water obtained from the source is distributed within the field. In general, the goal is to supply the entire field uniformly with water, so that each plant has the amount of water it needs, neither too much nor too little.The modern methods are efficient enough to achieve this goal.

Surface

In surface irrigation systems, water moves over and across the land by simple gravity flow in order to wet it and to infiltrate into the soil. Surface irrigation can be subdivided into furrow,borderstrip or basin irrigation. It is often called flood irrigation when the irrigation results in flooding or near flooding of the cultivated land. Historically, this has been the most common method of irrigating agricultural land.

Where water levels from the irrigation source permit, the levels are controlled by dikes, usually plugged by soil. This is often seen in terraced rice fields (rice paddies), where the method is used to flood or control the level of water in each distinct field. In some cases, the water is pumped, or lifted by human or animal power to the level of the land!

Localized

Localized irrigation is a system where water is distributed under low pressure through a piped network, in a pre-determined pattern, and applied as a small discharge to each plant or adjacent to it. Drip irrigation, spray or micro-sprinkler irrigation and bubbler irrigation belong to this category of irrigation methods.

Drip

Drip irrigation, also known as trickle irrigation, functions as its name suggests. in this system water falls drop by drop just at the position of roots.Water is delivered at or near the root zone of plants, drop by drop. This method can be the most water-efficient method of irrigation, if managed properly, since evaporation and runoff are minimized. In modern agriculture, drip irrigation is often combined with plastic mulch, further reducing evaporation, and is also the means of delivery of fertilizer. The process is known as fertigation.

Deep percolation, where water moves below the root zone, can occur if a drip system is operated for too long or if the delivery rate is too high. Drip irrigation methods range from very high-tech and computerized to low-tech and labor-intensive. Lower water pressures are usually needed than for most other types of systems, with the exception of low energy center pivot systems and surface irrigation systems, and the system can be designed for uniformity throughout a field or for precise water delivery to individual plants in a landscape containing a mix of plant species. Although it is difficult to regulate pressure on steep slopes, pressure compensating emitters are available, so the field does not have to be level. High-tech solutions involve precisely calibrated emitters located along lines of tubing that extend from a computerized set of valves.

Sprinkler

In sprinkler or overhead irrigation, water is piped to one or more central locations within the field and distributed by overhead high-pressure sprinklers or guns. A system utilizing sprinklers, sprays, or guns mounted overhead on permanently installed risers is often referred to as a solid-set irrigation system. Higher pressure sprinklers that rotate are called rotors and are driven by a ball drive, gear drive, or impact mechanism. Rotors can be designed to rotate in a full or partial circle. Guns are similar to rotors, except that they generally operate at very high pressures of 40 to 130 lbf/in^2 (275 to 900 kPa) and flows of 50 to 1200 US gal/min (3 to 76 L/s), usually with nozzle diameters in the range of 0.5 to 1.9 inches (10 to 50 mm). Guns are used not only for irrigation, but also for industrial applications such as dust suppression and logging.

Sprinklers can also be mounted on moving platforms connected to the water source by a hose. Automatically moving wheeled systems known as traveling sprinklers may irrigate areas such as small farms, sports fields, parks, pastures, and cemeteries unattended. Most of these utilize a length of polyethylene tubing wound on a steel drum. As the tubing is wound on the drum powered by the irrigation water or a small gas engine, the sprinkler is pulled across the field. When the sprinkler arrives back at the reel the system shuts off. This type of system is known to most people as a "waterreel" traveling irrigation sprinkler and they are used extensively for dust suppression, irrigation, and land application of waste water. Other travelers use a flat rubber hose that is dragged along behind while the sprinkler platform is pulled by a cable. These cable-type travelers are definitely old technology and their use is limited in today's modern irrigation projects.

Center pivot

Center pivot irrigation is a form of sprinkler irrigation consisting of several segments of pipe (usually galvanized steel or aluminum) joined together and supported by trusses, mounted on wheeled towers with sprinklers positioned along its length. The system moves in a circular pattern and is fed with water from the pivot point at the center of the arc. These systems are found and used in all parts of the world and allow irrigation of all types of terrain. Newer systems have drop sprinkler heads as shown in the image that follows.

Most center pivot systems now have drops hanging from a u-shaped pipe attached at the top of the pipe with sprinkler heads that are positioned a few feet (at most) above the crop, thus limiting evaporative losses. Drops can also be used with drag hoses or bubblers that deposit the water directly on the ground between crops. Crops are often planted in a circle to conform to the center pivot. This type of system is known as LEPA (Low Energy Precision Application). Originally, most center pivots were water powered. These were replaced by hydraulic systems (T-L Irrigation) and electric motor driven systems (Reinke, Valley, Zimmatic). Many modern sprinklers features GPSdevices.

Lateral move (side roll, wheel line)

A series of pipes, each with a wheel of about 1.5 m diameter permanently affixed to its midpoint and sprinklers along its length, are coupled together at one edge of a field. Water is supplied at one end using a large hose. After sufficient water has been applied, the hose is removed and the remaining assembly rotated either by hand or with a purpose-built mechanism, so that the sprinklers move 10 m across the field. The hose is reconnected. The process is repeated until the opposite edge of the field is reached. This system is less expensive to install than a center pivot, but much more labor intensive to operate, and it is limited in the amount of water it can carry. Most systems utilize 4 or 5-inch (130 mm) diameter aluminum pipe. One feature of a lateral move system is that it consists of sections that can be easily disconnected. They are most often used for small or oddly shaped fields, such as those found in hilly or mountainous regions, or in regions where labor is inexpensive.

Sub-irrigation

Subirrigation also sometimes called seepage irrigation has been used for many years in field crops in areas with high water tables. It is a method of artificially raising the water table to allow the soil to bemoistened from below the plants' root zone. Often those systems are located on permanent grasslands in lowlands or river valleys and combined with drainage infrastructure. A system of pumping stations, canals, weirs and gates allows it to increase or decrease the water level in a network of ditches and thereby control the water table.

Sub-irrigation is also used in commercial greenhouse production, usually for potted plants. Water is delivered from below, absorbed upwards, and the excess collected for recycling. Typically, a solution of water and nutrients floods a container or flows through a trough for a short period of time, 10-20 minutes, and is then pumped back into a holding tank for reuse. Sub-irrigation in greenhouses requires fairly sophisticated, expensive equipment and management. Advantages are water and nutrient conservation, and labor-saving through lowered system maintenance and automation. It is similar in principle and action to subsurface drip irrigation.

How an in-ground irrigation system works

Most commercial and residential irrigation systems are "in ground" systems, which means that everything is buried in the ground. With the pipes, sprinklers, emitters (drippers), and irrigation valvesbeing hidden, it makes for a cleaner, more presentable landscape without garden hoses or other items having to be moved around manually. This does, however, create some drawbacks in the maintenance of a completely buried system.

Controllers, zones, and valves

Most irrigation systems are divided into zones. A zone is a single irrigation valve and one or a group of drippers or sprinklers that are connected by pipes or tubes. Irrigation systems are divided into zones because there is usually not enough pressure and available flow to run sprinklers for an entire yard or sports field at once. Each zone has a solenoid valve on it that is controlled via wire by an irrigation controller. The irrigation controller is either a mechanical (now the "dinosaur" type) or electrical device that signals a zone to turn on at a specific time and keeps it on for a specified amount of time. "Smart Controller" is a recent term used to describe a controller that is capable of adjusting the watering time by itself in response to current environmental conditions. The smart controller determines current conditions by means of historic weather data for the local area, a soil moisture sensors (water potential or water content), rain sensor, or in more sophisticated systems satellite feed weather station, or a combination of these.

Emitters and sprinklers

When a zone comes on, the water flows through the lateral lines and ultimately ends up at the irrigation emitter (drip) or sprinkler heads. Many sprinklers have pipe thread inlets on the bottom of them which allows a fitting and the pipe to be attached to them. The sprinklers are usually installed with the top of the head flush with the ground surface. When the water is pressurized, the head will pop up out of the ground and water the desired area until the valve closes and shuts off that zone. Once there is no more water pressure in the lateral line, the sprinkler head will retract back into the ground. Emitters are generally laid on the soil surface or buried a few inches to reduce evaporation losses.

Problems in Irrigation

Irrigation can lead to a number of problems:

- Competition for surface water rights.
- Depletion of underground aquifers.
- Ground subsidence (e.g. New Orleans, Louisiana)
- Underirrigation or irrigation giving only just enough water for the plant (e.g. in drip line irrigation) gives poor soil salinity control which leads to increased soil salinity with consequent build up of toxic salts on soil surface in areas with high evaporation. This requires either leaching to remove these salts and a method of drainage to carry the salts away. When using drip lines, the leaching is best done regularly at certain intervals (with only

a slight excess of water), so that the salt is flushed back under the plant's roots.

- Overirrigation because of poor distribution uniformity or management wastes water, chemicals, and may lead to water pollution.
- Deep drainage (from over-irrigation) may result in rising water tables which in some instances will lead to problems of irrigation salinity requiring watertable control by some form of subsurface land drainage.
- Irrigation with saline or high-sodium water may damage soil structure owing to the formation of alkaline soil.

Present Extent

In the middle of the 20th century, the advent of diesel and electric motors led for the first time to systems that could pump groundwater out of major aquifers faster than it was recharged. This can lead to permanent loss of aquifer capacity, decreased water quality, ground subsidence, and other problems. The future of food production in such areas as the North China Plain, the Punjab, and the Great Plains of the US is threatened.

At the global scale, 2,788,000 km^2 (689 million acres) of agricultural land was equipped with irrigation infrastructure around the year 2000. About 68% of the area equipped for irrigation is located in Asia, 17% in America, 9% in Europe, 5% in Africa and 1% in Oceania. The largest contiguous areas of high irrigation density are found in North India and Pakistan along the rivers Ganges and Indus, in the Hai He, Huang He and Yangtze basins in China, along the Nile river in Egypt and Sudan, in the Mississippi-Missouri river basin and in parts of California. Smaller irrigation areas are spread across almost all populated parts of the world.

Generally about 80% of the country receives less than 1000mm of seasonal unreliable rainfall. Arable land area is estimated to be 44 million hectares. According to the National Irrigation Master Plan (NIMP) study done in 2002, it was revealed that of the territorial area, potential irrigable area is estimated at 29.4 million hectares. Out of the total potential irrigable area, 2.3 million hectares have high potential, 4.8 million hectares have medium potential, and 22.3 million hectares are of low potential for irrigation development respectively. Current cultivated area is estimated to be 10.1 million hectares or 23% of the arable land, of which only about 2.3% is currently under irrigation. About 80% of this area is under smallholder and traditional irrigation and yet irrigation accounts for 8% of the agricultural production in the country of mainly rice, maize and vegetables.

The government acknowledges that, though traditional irrigation is potentially the most productive sub-sector of irrigation it has not been given adequate attention in the past. Traditional irrigation is, as a result, characterized by the several constraints. In Kilimanjaro region, for example, it is estimated that 52,000ha of land in Kilimanjaro are under traditional Irrigation system mainly in the highlands with 508 furrows covering an area of about 16,647ha. Traditional irrigation systems operate below capacity due to inadequate water supply caused by poor irrigation infrastructure and lack of skills in water management and poor farming practices. About 20% of the furrows built in the 1950s and 1960s need rehabilitation to improve irrigation efficiencies and productivity.

Majority of smallholder irrigators, who are the main crop growers and livestock keepers do not have land title deeds but hold land under deemed right of occupancy. Farmers consider this as a disincentive for investments in land management.

There has been a serious problem of the government trying to focus on expensive capital-intensive irrigation technologies. In traditional irrigation, expensive techniques are used instead of the low-cost innovative technologies (pipe conveyance, drip irrigation, treadle pumps, and water harvesting) affordable by most small irrigators.

Providing Solutions

TIP provides technical assistance to water user groups with regard to the improvement of existing irrigation system. The decision to implement proposed solutions is made in a participatory manner with target groups to be technically feasible, sustainable and cost effective.

Improvement of traditional irrigation systems is not limited to construction structures. Village Technicians are trained to operate and maintain an irrigation system. This training enables the groups to manage improved irrigation systems sustainability. As the farmers' groups become strong, they operate communal funds from annual fees and penalty fines that are used for maintenance costs. Male and female farmers are also trained in Water Management in order to make a more efficient use for the most limited natural resource "Water".

POLICY ISSUES IN IRRIGATED AGRICULTURE

Irrigation is an essential component of sustainable agricultural development but it is not a unique sector, since it faces challenges similar to those confronting other public and private sector economic activities. The previous two sections (II and III) of this report examine how various demand-side policy measures can help shape decisions that encourage water-use efficiency. However, while appropriate policies and regulations are necessary for improved water productivity, a variety of additional water-saving measures are required in the irrigation sector.

Some water-saving measures involve taking more advantage of the scientific, engineering and technological advances in soils, plants and irrigation. Other measures focus on administrative and managerial reforms to improve efficiency, including the decentralization of public irrigation agencies and a greater reliance on farmer-owned and farmer-operated irrigation.

This final section highlights three key irrigation issues: declining growth and investment trends in irrigation; the difficulties imposed by irrigation-induced environmental degradation; and efforts to reform managerial and administrative systems. Many of today's irrigation-related problems appear imposing and even overwhelming. The purpose of this section is not to present a discouraging overview of irrigation's future but to focus on important issues that will shape its future. When scarce water is under human control in irrigation systems, there are many opportunities to use it optimally. Understanding the problems associated with improper irrigation as well as the potential for efficient irrigation is a first step in the search for these opportunities.

Irrigation in the 1990s and beyond

Many pressing irrigation issues reflect the various economic, social and political influences affecting society in this period of changing economic conditions and environmental concerns. At times, irrigation-related problems are the result of a distorted macroeconomy which, despite providing operating subsidies, renders farming unprofitable and results in repeated under investments on farms over long periods. If the macroeconomy is performing poorly and if prices in the economy are markedly distorted from their real value to society, then it is inevitable that irrigation will suffer.

In other cases, irrigation itself is the problem. The overall performance of many irrigation projects is disappointing. Evaluations document a wide range of problems, including: cost and time overruns; poor management; the non-realization of full, planned benefits; adverse environmental and health impacts; and the exacerbation of inequities in the existing social and economic distribution of assets among farmers.

Trends in irrigated area

In 1800, approximately 8 million ha of farmland were irrigated around the world. By the end of the nineteenth century, the irrigated area had expanded to 48 million ha, mostly as a result of large water projects in India and what is now Pakistan. By 1990, net irrigated area reached 237 million ha, of which nearly three-fourths were in the developing countries. China, India and Pakistan alone now account for about 45 percent of the world's irrigated area and 60 percent of the developing country total.

On a global basis, irrigated area expanded by an average of 1 percent per year during the early 1960s, reaching a maximum annual rate of 2.3 percent from 1972 to 1975. The rate of expansion began decreasing after 1975 and is now less than 1 percent per year. At current levels of population growth, the slower expansion in irrigated areas is resulting in an unprecedented decline in the per caput amount of irrigated land.

Among the reasons for the decrease are increased construction costs, falling real prices for wheat and rice, a growing awareness of environmental and social costs and poor irrigation performance at the farm and project levels.

Crop prices and construction costs

Construction and operation costs for irrigation projects have risen steadily over the past four decades as the world's best land and most of the readily available water supplies have been developed. Over the same time period, world cereal prices fell sharply.

For example, the real price of rice fell by 40 percent between the mid-1960s and the late 1980s.

In Indonesia and India, the real costs of irrigation have doubled since the early 1970s. In the Philippines and Thailand, irrigation construction costs have increased by 50 percent and in Sri Lanka, costs have tripled. After reviewing various recent reports and evaluations, including many from the World Bank and FAO, Postel reports that: "Today, capital costs for new irrigation capacity run between $1 500 and $4 000 per ha for large projects in China, India, Indonesia, Pakistan, the Philippines and Thailand. They climb toward $6 000 per ha in Mexico. In Africa, where roads and other infrastructure are often lacking and parcels that can be irrigated are relatively small, per hectare costs have climbed to $10 000-$20 000, and sometimes even higher. Not even double-cropping of higher-valued crops can make irrigation systems at the top end of this spectrum economical." It is not just large-scale projects that have become so expensive; FAO estimates that even medium-sized irrigation construction costs range from $2 400 per ha in Asia to $7 200 in Africa.

Likewise, the modernization of existing irrigation projects is becoming increasingly expensive. Many old projects designed for monocropping have produced gradually declining yields. These systems need to be redesigned to permit crop diversity, increase yields, conserve water and reduce environmental hazards. Modernization involves canal lining, improved hydraulic control structures, better land development and appropriate irrigation techniques.

Increasing real investment costs (including recognition of environmental costs) and falling crop prices have resulted in a substantial reduction in new irrigation projects. Aggregate lending for irrigation by both international donors and the major cereal-producing countries in the 1980s was only half what it was in the 1970s. This reduced investment has, in turn, resulted in falling growth rates in irrigated area.

Irrigation and land degradation

Escalating demand pressure on water resources during the past decade is only one of the concerns for policy-makers. Both watersheds supporting water resources and the land base supporting irrigated agriculture are becoming degraded. Industrial and domestic pollution are affecting irrigated agriculture, while sedimentation washed from denuded hillsides is filling up irrigation reservoirs and small tanks. At the same time, improper irrigation practices are causing waterlogging, salinization, soil erosion and water pollution which directly affects other irrigators.

FAO estimates that of the 237 million ha currently irrigated, about 30 million ha are severely affected by salinity and an additional 60 million to 80 million ha are affected to some extent. UNEP recently reported that the rate of loss of irrigated land from waterlogging and salinity is 1.5 million ha per year. Millions of hectares of irrigated land, from Morocco to Bangladesh and from northwestern China to central Asia, suffer from this progressive condition. Salinity-affected areas as a percentage of total irrigated area is estimated to be 10 percent in Mexico, 11 percent in India, 21 percent in Pakistan, 23 percent in China and 28 percent in the United States.

Salinity is caused by a combination of poor drainage and high evaporation rates which concentrate salts on irrigated land; it mainly occurs in arid and semi-arid regions. Even good-quality irrigation water contains some dissolved salt and can leave behind tonnes of salt per hectare each year. Unless this salt is washed down below the root level, soil salinity will result. A number of factors influence salinity, including the water table depth, the capillary characteristics of the soil and management practices regarding the amount of water applied in excess of plant evapotranspiration to leach the salts.

A related concern is the rapid rise in groundwater levels, leading to waterlogging and depressed crop yields. Waterlogging is not an inevitable result of irrigation. It occurs when excessive water is used in systems with finite natural drainage capacities. Seepage occurs if: soils are very light; canals and watercourses are not lined or maintained; farmers near the head of a system withdraw or apply excessive amounts of water; fields are not levelled; and/or the delivery system cannot respond to rainfall by closing inflows.

Irrigation can raise groundwater levels to within a metre or so of the surface, which results in secondary salinization when the groundwater brings to the surface dissolved salts from the aquifer, subsoil and root zone. If seepage and horizontal recharge exceed evaporation and natural drainage, then groundwater levels rise, eventually causing waterlogging. In arid areas where the upward movement of water and evaporation exceeds downward percolation and where the groundwater, soil or irrigation water contains some salt, the buildup of salt in the soil surface layers will eventually reach toxic levels.

Countries suffering from waterlogging and salinity face a dilemma. They cannot force abandonment of affected lands because of the growing populations that depend on them, but neither can they afford to drain the lands. The cure for rising water tables is drainage and improved water management to reduce percolation, but drainage is expensive and improved water management requires both on-farm investment and the training of extension personnel and farmers.

At the end of the 1950s, Pakistan began to sink wells to pump out salt-laden water. The initial capital costs were high and operating costs have been increasing constantly. Between 1971 and 1985 the cost of managing the wells and drains rose fivefold. Today, Pakistan spends more on reclaiming land than on irrigation. The cost of maintaining drains is five times the recurrent costs of supplying water and most farmers pay only half the costs of delivering irrigation.

Some policy-makers argue that, as the value of water rises, management will improve and much of the wastage (which creates unhealthy conditions as well as increased drainage requirements) will be reduced. Farmers may then be willing to pay for a drainage service that makes their investments sustainable.

Irrigation: good government and good management

Political pressures often prevent the implementation of apparently sensible and fair reforms for water services. Where water is regarded as a special commodity or has emotional or religious importance, governments are reluctant to charge farmers for irrigation. Policy-makers often find it extremely difficult to raise sufficient revenue to match even their priority needs. The practical effects on traditional public service activities may be harsh. Water has been one of the first sectors to feel the effects of budget-saving efforts and changing resource availability. However, it is unlikely to be treated more austerely than other areas. Box 19 discusses the relationship between good government and irrigation performance.

A prime opportunity for irrigation advancement - and indeed for development in general - lies in the enormous potential of the 237 million ha already operating. While the total value of irrigation investment in the developing world today is about $1 000 billion, the returns are well below the known potential. Many irrigation schemes need substantial investment to be completed, modernized or expanded. Although it is increasingly expensive, rehabilitation can yield high returns.

Good Government and Irrigation Performance

Improvement in irrigation performance depends on good government, or governance. This may be an obvious assertion, but what exactly does the term mean for irrigation? There are four main elements of governance which can be considered at the national or the local level: the legitimacy of government; itsaccountability, its competence; and its respect for human rights and the rule of law.

Legitimacy refers to the way in which a population gives consent to be governed, how they are consulted and whether the consent can be withdrawn. Accountability of politicians and officials is tested by how they explain their role and decisions, provide information and can be held responsible for their behaviour. A government demonstrates competence in formulating policies and translating them into action in a timely and effective way. Governments who respect human rights establish a framework of known laws, applicable to all, without bias or corruption, with limits on and protection against the exercise of arbitrary power. We can illustrate how these four elements of governance might affect irrigation with simple examples.

Legitimacy. When a new project is planned, are those living in the area consulted about the design of the scheme? Are there recognized representative groups of farmers, including women? Are the office holders elected and accountable to the members? Do these groups participate in decisions that affect them? For example, if a groundwater irrigation project threatens the availability of drinking-water from hand pumps, are self-help groups informed and invited to make representations?

Accountability. Are the financial plans of the irrigation scheme made public and arrangements made to explain them to farmers? Are there performance criteria with audit arrangements to ensure that officials adhere to the rules and, if they fail to perform satisfactorily, call them to account? Are officials responsive?

Competence. Can the professional staff prepare accurate budgets and effectively deliver services such as timely canal maintenance? Are there arrangements for training them or replacing them with competent officers if they fall short of their duties?

Rule of law. Is there a clear legal framework to regulate groundwater abstraction to prevent overpumping of the aquifers? Is it enforced? Can pollution by industry or by saline water from upstream drainage projects be regulated? Are illegal extractions by farmers at the head of canals monitored and offenders charged by legal processes that are fair, timely, objective and without discrimination on grounds of race, gender or minority status?

Beyond capital investment, investment in human resources is needed to improve economy, efficiency and effectiveness in management.

Investment in irrigation management can be an integral component of a productive rural employment policy and resource management can be improved to match developments in engineering and science. Those concerned with irrigation development and the welfare of farmers and their families are beginning to take a broader view of their responsibilities, paying as much attention to management and development as to engineering.

A review of recent literature on public sector policy and management suggests the need for transforming the typical irrigation authority from a technical, engineering and construction culture to an organization that is focused more on people (farmers) and that is more autonomous and more responsive to customers, i.e. a strategic management service agency.

In the developing world, many water subsector groups recognize their role in the management of valuable water and other complementary resources for economic growth and poverty alleviation.

Irrigation management does not generally have a good record, but the full extent of the problems has been obscured by a lack of performance assessment criteria and a variety of hidden subsidies. It has often been assumed that, if farmers are not complaining, then management must be good. A lack of complaints in a highly subsidized system is not a good indicator of efficiency. Similarly, managers with obvious problems have sometimes had to cope as best they could with a range of exogenous factors that limit their effectiveness, for instance overvalued exchange rates.

One example of a remarkably successful administrative reform is found in the Philippines, where irrigation operating expenses have been reduced, revenue has matched operating costs and operational performance has improved in the process. Box 20 details the Philippines' experience, demonstrating what can be achieved with good planning, dedication, determined technical staff and political support.

Irrigation management: water user associations and NGOs

An increasing number of private sector groups, including water user associations and other NGOs, are taking over some public sector irrigation responsibilities. The inclusion of water users in irrigation planning, management and ownership is proving to be an effective method for increasing irrigation system efficiency in many cases.

Studies throughout the world demonstrate that user participation in irrigation services improves access to information, reduces monitoring costs, establishes a sense of ownership among farmers and increases transparency as well as accountability in decision-making.

Water user associations are expected to increase in number and importance over the next decade as the stress on self-reliance increases. Already, governments are turning many aspects of public irrigation systems over to water user associations. Well-documented examples can be seen in Argentina, Colombia, Indonesia,

Mexico, Nepal, the Philippines, Sri Lanka and Tunisia. In Indonesia, for example, the government had transferred more than 400 irrigation systems, covering 34 000 ha, to water user associations by 1992.62 In the future, as farmer financing becomes more commonplace, user groups will become even more powerful.

Other NGOs undertake a wide range of water-related functions, from developing projects for rural water supplies and minor irrigation to fostering water user associations for water management purposes. Some NGOs encourage farmers to try new technologies, for example the catchment protection and sprinkler irrigation techniques introduced by the Aga Khan Rural Support Programme in Gujarat, India.

Many NGOs stem from local initiatives and operate as independently funded and self-managed groups. These organizations bring fresh views, new ideas and participatory working methods to other areas of development policy and practice. Much of their success is attributed to their local knowledge as well as their interest in and experience of regional conditions. They have been particularly active in promoting the interests of poor and disadvantaged groups through articulate and forceful advocacy and service provision. In addition, the local base of NGOs may allow them to reach vulnerable or remote groups which are exceptionally difficult to reach with conventional public schemes.

Water is not an easy sector in which to promote cooperation, but the potential gains are high, which makes renewed efforts worthwhile. Resolution of many water allocation and development problems requires a common willingness to forego personal benefit for the social good. Government efforts to promote personal sacrifice through economic policies, laws and regulations that require self-restraint, such as water rationing or optimum groundwater pumping regimes, have seldom proved effective. On the other hand, with their close local contacts and skills in group mobilization and cohesion, NGOs can provide the institutional leadership required to bring about socially optimum solutions.

Future directions in water management policy

Sustainable agricultural development depends on sustainable water use. Governments today recognize that the search for sustainable economic growth requires, in part, both economy-wide and sector-specific policy reforms. Economy-wide policies attempt to create a favourable macroeconomic environment while water sector policies, for example, seek to encourage resource efficiency among water users. The current emphasis on macro-economic policy reforms and economic liberalization has several important implications for irrigation. One of the most important is that the era of large direct and indirect subsidies is nearly over. Moreover, recognition of the value of water (and the high cost of turning a water source into a service delivered to a farm) makes the water sector a prime target for further policy reforms. Nonetheless, irrigation remains a resource-hungry sector in this transitional period. Even successful irrigation consumes large quantities of capital and foreign exchange and ties up scarce skilled personnel.

Like many public sector personnel, irrigation managers must walk a fine line between a tighter control of finance, the need for more positive active leadership and better planning of resource allocations, on the one side, and the contradictory need for more ideas from below (farmer customers), on the other. Financial pressures are likely to be the dominant influence. Irrigation as a public sector agency still relies on budget allocations to obtain financing. Many argue that this gives little incentive to save money and may, in fact, have the reverse effect.

As private sector disciplines are applied in irrigation, policy-makers are finding that: agencies become more supportive of farmers' own efforts and less inclined to make all key decisions before informing farmers accordingly; management seeks more consensus on priorities, more information about the basis of decisions and a common view of external factors affecting management; irrigation schemes seek and receive more autonomy; the financial responsibilities and accountability of managers increases; and managers shift focus from their ministries and governments, depending on the amount of finance generated by service fees.

At present, water engineering feeds off mathematics, physics, chemistry and biology; management feeds off economics, psychology, political science, history and philosophy and also has a high requirement for personal skills in communication, negotiation and teamwork. However, as changes occur in irrigation in many countries, managers find they must acquire new knowledge and a new set of skills.

7

RAIN WATER HARVESTING

Rainwater harvesting is the accumulating and storing, of rainwater for reuse, before it reaches the aquifer. It has been used to provide drinking water, water for livestock, water for irrigation, as well as other typical uses given to water. Rainwater collected from the roofs of houses, tents and local institutions, can make an important contribution to the availability of drinking water.It can supplement the sub soil water level and increase urban greenery. Water collected from the ground, sometimes from areas which are especially prepared for this purpose, is called Storm water harvesting. In some cases, rainwater may be the only available, or economical, water source. Rainwater harvesting systems can be simple to construct from inexpensive local materials, and are potentially successful in most habitable locations. Roof rainwater can't be of good quality and may require treatment before consumption. As rainwater rushes from your roof it may carry pollutants in it such as the tiniest bit of mercury from coal burning buildings to bird feces. Although some rooftop materials may produce rainwater that is harmful to human health, it can be useful in flushing toilets, washing clothes, watering the garden and washing cars; these uses alone halve the amount of water used by a typical home. Household rainfall catchment systems are appropriate in areas with an average rainfall greater than 200 mm (7.9 in) per year, and no other accessible water sources. Overflow from rainwater harvesting tank systems can be used to refill aquifers in a process called groundwater recharge, though this is a related process, it must not be confused with Rainwater harvesting.

There are a number of types of systems to harvest rainwater ranging from very simple to the complex industrial systems. The rate at which water can be collected from either system is dependent on the plan area of the system, its efficiency, and the intensity of rainfall (i.e. annual precipitation (mm per annum) x square meter of catchment area = litres per annum yield) ... a 200 square meter roof catchment catching 1,000mm PA yields 200 kLPA.

BASIC CONFIGURATION

Rainwater is harvesting systems channel rainwater that falls on to a roof into storage via a system of gutters and pipes. The first flush of rainwater after a dry season should be allowed to run to waste as it will be contaminated with dust, bird droppings etc. Roof gutters should have enough, and large enough to carry peak flows. Storage tanks should be covered to prevent mosquito breeding and to reduce evaporation losses, contamination and algal growth. Rainwater harvesting systems require regular maintenance and cleaning to keep the system hygienic.

Subsurface dyke

A subsurface dyke is built in an aquifer to obstruct the natural flow of groundwater, thereby raising the groundwater level and increasing the amount of water stored in the aquifer. The subsurface dyke at Krishi Vigyan Kendra, Kannur under Kerala Agricultural University with the support of ICAR, has become an effective method for ground water conservation by means of rain water harvesting technologies. The sub-surface dyke has demonstrated that it is a feasible method for conserving and exploiting the groundwater resources of the Kerala state of India. The dyke is now the largest rainwater harvesting system in that region.

Groundwater recharge

Rainwater may also be used for groundwater recharge, where the runoff on the ground is collected and allowed to be absorbed, adding to the groundwater. In the US, rooftop rainwater is collected and stored in sump.

Advantages in urban areas

Rainwater harvesting can (a) assure an independent water supply during water restrictions, that is though somewhat dependent on end use and maintenance, (b) usually of acceptable quality for household needs and (c) renewable at acceptable volumes despite forecast climate change (CSIRO, 2003). It produces beneficial externalities by reducing peak storm water runoff and processing costs. In municipalities with combined sewer systems, reducing storm runoff is especially important, because excess runoff during heavy storms leads to the discharge of raw sewage from outfalls when treatment plant capacity cannot handle the combined flow. RH systems are simple to install and operate. Running costs are negligible, and they provide water at the point of consumption. (d) Rainwater harvesting in urban communities has been made possible by companies such as Living Rainwater

Tanks . Tanks such as this provide an attractive yet effective solution to rainwater catchment.

Rain water harvesting law Some states have water law in favor of land owners, and some states own all water rights, leaving you only water rights they are willing to grant through permits. For example, in Colorado you may not catch, collect or harvest rain water from your roof unless you first buy a permit.

Quality

As rainwater may be contaminated due to pollutants like microscopic germs etc., it is often not considered suitable for drinking without treatment. However, there are many examples of rainwater being used for all purposes - including drinking - following suitable treatment.

Rainwater harvested from roofs can contain human, animal and bird faeces, mosses and lichens, windblown dust, particulates from urban pollution, pesticides, and inorganic ions from the sea (Ca, Mg,Na, K, Cl, SO4), and dissolved gases (CO2, NOx, SOx). High levels of pesticide have been found in rainwater in Europe with the highest concentrations occurring in the first rain immediately after a dry spell; the concentration of these and other contaminants are reduced significantly by diverting the initial flow of water to waste as described above. The water may need to be analysed properly, and used in a way appropriate to its safety. In the Gansu province for example, harvested rainwater is boiled in parabolic solar cookers before being used for drinking. In Brazil alum and chlorine is added to disinfect water before consumption. So-called "appropriate technology" methods, such as solar water disinfection, provide low-cost disinfection options for treatment of stored rainwater for drinking.

System sizing

It is important that the system is sized to meet the water demand throughout the dry season. Generally speaking, the size of the storage tank should be big enough to meet the daily water requirement throughout the dry season. In addition, the size of the catchment area or roof should be large enough to fill the tank..

In the present scenario management and distribution of water has become centralized. People depend on government system, which has resulted in disruption of community participation in water management and collapse of traditional water harvesting system.

As the water crisis continues to become severe, there is a dire need of reform in water management system and revival of traditional systems. Scientific and technological studies needs to be carried out to assess present status so as to suggest suitable mitigative measures for the revival to traditional system/wisdom.

Revival process should necessarily be backed by people's initiative and active public participation.

Living creatures of the universe are made of five basic elements, viz., Earth, Water, Fire, Air and Sky, Obviously, water is one of the most important elements and no creature can survive without it. Despite having a great regard for water, we seem to have failed to address this sector seriously. Human being could not save and conserve water and it sources, probably because of its availability in abundance. But this irresponsible attitude resulted in deterioration of water bodies with respect to quantity and quality both. Now, situation has arrived when even a single drop of water matters. However. " better late than never", we have not realized the seriousness of this issue and initiated efforts to overcome those problems.

COMPONENTS OF A RAINWATER HARVESTING SYSTEM

A rainwater harvesting system comprises components of various stages - transporting rainwater through pipes or drains, filtration, and storage in tanks for reuse or recharge. The common components of a rainwater harvesting system involved in these stages are illustrated here.

1. Catchments: The catchment of a water harvesting system is the surface which directly receives the rainfall and provides water to the system. It can be a paved area like a terrace or courtyard of a building, or an unpaved area like a lawn or open ground. A roof made of reinforced cement concrete (RCC), galvanised iron or corrugated sheets can also be used for water harvesting.
2. Coarse mesh at the roof to prevent the passage of debris
3. Gutters: Channels all around the edge of a sloping roof to collect and transport rainwater to the storage tank. Gutters can be semi-circular or rectangular and could be made using:

"Locally available material such as plain galvanised iron sheet (20 to 22 gauge), folded to required shapes.

"Semi-circular gutters of PVC material can be readily prepared by cutting those pipes into two equal semi-circular channels.

"Bamboo or betel trunks cut vertically in half.

The size of the gutter should be according to the flow during the highest intensity rain. It is advisable to make them 10 to 15 per cent oversize.

Gutters need to be supported so they do not sag or fall off when loaded with water. The way in which gutters are fixed depends on the construction of the house; it is possible to fix iron or timber brackets into the walls, but for houses having wider eaves, some method of attachment to the rafters is necessary.

Conduits: Conduits are pipelines or drains that carry rainwater from the catchment or rooftop area to the harvesting system. Conduits can be of any material like polyvinyl chloride (PVC) or galvanized iron (GI), materials that are commonly available.

First-flushing: A first flush device is a valve that ensures that runoff from the first spell of rain is flushed out and does not enter the system. This needs to be done since the first spell of rain carries a relatively larger amount of pollutants from the air and catchment surface.

Filter : The filter is used to remove suspended pollutants from rainwater collected over roof. A filter unit is a chamber filled with filtering media such as fibre, coarse sand and gravel layers to remove debris and dirt from water before it enters the storage tank or recharge structure. Charcoal can be added for additional filtration.

(i) Charcoal water filter : A simple charcoal filter can be made in a drum or an earthen pot. The filter is made of gravel, sand and charcoal, all of which are easily available.

(ii) Sand filters : Sand filters have commonly available sand as filter media. Sand filters are easy and inexpensive to construct. These filters can be employed for treatment of water to effectively remove turbidity (suspended particles like silt and clay), colour and microorganisms.

In a simple sand filter that can be constructed domestically, the top layer comprises coarse sand followed by a 5-10 mm layer of gravel followed by another 5-25 cm layer of gravel and boulders.

(a) Dewas filters : Most residents in Dewas, Madhya Pradesh, have wells in their houses. Formerly, all that those wells would do was extract groundwater. But then, the district administration of Dewas initiated a groundwater recharge scheme. The rooftop water was collected and allowed to pass through a filter system called the Dewas fillter, designed by Mohan Rao , district collecter of Dewas, and engineers of the rural engineering services. The water thus filtered is put into the service tubewell. The filter consists of a polyvinyl chloride (PVC) pipe 140 mm in diameter and 1.2m long. There are three chambers. The first purification chamber has pebbles varying between 2-6 mm, the second chamber has slightly larger pebbles, between 6 and 12 mm and the third chamber has the largest - 12-20 mm pebbles. There is a mesh at the outflow side through which clean water flows out after passing through the three chambers. The cost of this filter unit is Rs 600.

Filter for large rooftops

When rainwater is harvested in a large rooftop area, the filtering system should accommodate the excess flow. A system is designed with three concentric circular chambers in which the outer chamber is filled with sand, the middle one with coarse aggregate and the inner-most layer with pebbles.

This way the area of filtration is increased for sand, in relation to coarse aggregate and pebbles. Rainwater reaches the centre core and is collected in the sump where it is treated with few tablets of chlorine and is made ready for consumption.

Rainwater Harvesting and Purification System

In January 1996 we installed a rainwater catchment system to capture Oregon's abundant rainfall. Portland receives between 3 and 4 feet of rainfall annually. During a gentle rain a typical Oregon downspout sheds several gallons per minute. Our twelve hundred square foot roof captures on average 3600 cubic feet (27,000 gallons) of water per year.

In 1998 we received approval from the city of Portland to use this water for all household use. This system, which cost less than $1,500, consists of the following components:

"A 1500 gallon plastic cistern, approximate cost: $500. Purchased from Northwest Irrigation, Tangent, Oregon, 541-928-0114. Contact local agriculture/ farm stores for best prices.

"A 1/2 horsepower shallow-well pump to pressurize the water to between 20 and 30 psi (pressure is adjustable), approximate cost: $250. I utilized a Jaccuzzi brand pump.

"Plastic (outdoor PVC and indoor CPVC) piping to connect to the household cold water system.

"Two particulate filters in series, rated at 20 and 5 micron particle size, approximate cost: $20 each; replaceable filter cartridges cost $3-5 each.

"An ultraviolet light sterilizer capable of sterilizing water at 10 gallons per minute. This appliance was recently approved for use in Oregon. I used the PURA (1-800-292-PURA, Valencia, California) model UV20-1, cost approximately $350. Uses about 40 watts. Fluorescent ultraviolet light rated at 9600 hours, about one year of continuous use. Replacement cost of fluorescent tube: about $80.

"Screen covering the cistern to prevent entry of mosquitoes and to catch any large particles that make it past the gutter screening.

"A roof-washer which wastes the first 7.5 gallons of captured water which has "washed" the roof. Once the roof washer has filled, the rest of the water flows to the cistern. See below for details.

" A 20 gallon water butyl rubber diaphragm pressure storage tank, approximate cost: $150.

"A reduced pressure backflow prevention device. This was required by the city to prevent flow of rainwater into the public system. Cost: $120. This would not be necessary if we used rainwater exclusively. However, Oregon has very dry summers and our cistern is exhausted by July. We currently depend on city water during the summer. The city requires annual inspection of these devices, costing about $30.

"A (optional) water meter to measure rainwater output, approximate cost: $45.

Maintenance consists of keeping gutters and cistern screen clean. Filters and ultra-violet lamp will need periodic replacement. The tank is thoroughly cleaned annually in the summer when it empties. Backflow prevention device requires annual inspection. Public health authorities recommend periodic testing of water for fecal coliform bacteria, as for any private water system. Several recent tests showed none. The inside components of our system, pictured here, take up about 6 square feet of floor space.

At the current time we continue to use the public water supply only for summertime water and occasional drinking and cooking. In fact, during the rainy season, which lasts from about September to June, our only connection to the public utility is one faucet at the kitchen sink which uses less than one gallon per day, which got us into hot water with the city water bureau.

In my research on rainwater catchment systems the best single reference I have come across for detailed design guidelines is the Texas Water Development Board's Texas Guide to Rainwater Harvesting.

Roofwashers. A simple prototype is shown in the TG. It consists of a length of pipe for storage of the initial flush of water with a trickle valve (hose bib just slightly opened) and clean out valve at the bottom. Only when the this pipe fills is water then allowed to continue into the cistern. It's very simple, no moving parts. The only thing I would change is to have a narrow section or trap configuration at the top to reduce mixing of the flush water with the still arriving (clean) water. Yet another method to aid this is to add a lightweight (like styrofoam) ball that would seal the intake when the roof washer fills. This simple design is very inexpensive, easy to drain or clean manually, and works very well. The TG suggests one gallon of washer capacity for each 100 square feet of roof. So make your roof washer pipe length long enough. For our model we used 20 feet of 3" ABS. We made it in the shape of a giant U to get this length. Remember, volume equals length times area. Area equals pi times radius squared (in our case 3 inches internal diameter, or .25 foot) and one cubic foot equals 7.5 gallons. To avoid long lengths of roofwasher pipe, it makes sense to use larger diameters. Portland's chief residential plumbing inspector commented that our use of ABS didn't conform to code as plastic may eventually decay in sunlight. Therefore, you should use copper, iron, or other sunlight-resistant materials to be completely correct.

Rainbarrels. A rainwater harvesting system can be as simple as a barrel connected to a downspout. Check the Rainbarrel Tutorial for tips on how to put together a system for as little as $15-20. One of our neighbors has connected his rainbarrel to his basement washing machine and gets virtually all his laundry water from this super-soft source for a miniscule investment.

One notable advantage of rainwater is its softness. Rainfall in the Portland area contains about 5 mg/liter of dissolved minerals. Compare this with some hard groundwater which exceeds 500 mg/liter. Portland city water, which has an exceptionally pure source, is rated at 18 mg/liter.

According to two officials in Alaska and Hawaii with whom I have communicated, there is a long established tradition of rainwater collection in some parts of their states. According to Sourcebook Harvested Rainwater, in some areas of the Caribbean, new houses are required to have rainwater capture systems. Hawaii apparently is currently developing (or has already developed) guidelines. In Oregon, there is no regulation of water quality for individual residences -- this is left up to the homeowner. The only regulations I have come across relating to rainwater harvesting are from Ohio, whose Department of Health Administrative Code regulates private water systems. Note, in particular, Rules 3701-28-09 Continuous disinfection and 3701-28-13 Construction and surface design of cisterns, hauled water storage tanks, and roof washers.

RAIN HARVESTING

Traditional Methods

Tankas

Tankas (small tank) are underground tanks, found traditionally in most Bikaner houses. They are built in the main house or in the courtyard. They were circular holes made in the ground, lined with fine polished lime, in which raiwater was collected. Tankas were often beautifully decorated with tiles, which helped to keep the water cool. The water was used only for drinking. If in any year there was less than normal rainfall and the tankas did not get filled, water from nearby wells and tanks would be obtained to fill the household tankas. In this way, the people of Bikaner were able to meet their water requirements. The tanka system is also to be found in the pilgrim town of Dwarka where it has been in existence for centuries. It continues to be used in residential areas, temples, dharamshalas and hotels.

Khadin

A khadin, also called a dhora, is an ingenious construction designed to harvest surface runoff water for agriculture. Its main feature is a very long (100-300 m) earthen embankment built across the lower hill slopes lying below gravelly uplands.

Sluices and spillways allow excess water to drain off. The khadinsystem is based on the principle of harvesting rainwater on farmland and subsequent use of this water-saturated land for crop production.

First designed by the Paliwal Brahmins of Jaisalmer, western Rajasthan in the 15th century, this system has great similarity with the irrigation methods of the people of Ur (present Iraq) around 4500 BC and later of the Nabateans in the Middle East. A similar system is also reported to have been practised 4,000 years ago in the Negev desert, and in southwestern Colorado 500 years ago.

Vav / vavdi / Baoli / Bavadi

Traditional stepwells are called vav or vavadi in Gujarat, or baolis or bavadisin Rajasthan and northern India. Built by the nobility usually for strategic and/or philanthropical reasons, they were secular structures from which everyone could draw water. Most of them are defunct today.

The construction of stepwells date from four periods: Pre-Solanki period (8th to 11th century CE); Solanki period (11th to 12th century CE); Vaghela period (mid-13th to end-14th century CE); and the Sultanate period (mid-13th to end-15th century CE).

Sculptures and inscriptions in stepwells demonstrate their importance to the traditional social and cultural lives of people.

Stepwell locations often suggested the way in which they would be used. When a stepwell was located within or at the edge of a village, it was mainly used for utilitarian purposes and as a cool place for social gatherings. When stepwells were located outside the village, on trade routes, they were often frequented as resting places. Many important stepwells are located on the major military and trade routes from Patan in the north to the sea coast of Saurashtra. When stepwells were used exclusively for irrigation, a sluice was constructed at the rim to receive the lifted water and lead it to a trough or pond, from where it ran through a drainage system and was channelled into the fields.

A major reason for the breakdown of this traditional system is the pressure of centralisation and agricultural intensification.

Ahar Pynes

This traditional floodwater harvesting system is indigenous to south Bihar.

In south Bihar, the terrain has a marked slope -- 1 m per km -- from south to north. The soil here is sandy and does not retain water. Groundwater levels are low. Rivers in this region swell only during the monsoon, but the water is swiftly carried away or percolates down into the sand. All these factors make floodwater harvesting the best option here, to which this system is admirably suited.

An ahar is a catchment basin embanked on three sides, the 'fourth' side being the natural gradient of the land itself. Ahar beds were also used to grow a rabi(winter) crop after draining out the excess water that remained after kharif(summer) cultivation.

Pynes are artificial channels constructed to utilise river water in agricultural fields. Starting out from the river, pynes meander through fields to end up in an ahar. Most pynes flow within 10 km of a river and their length is not more than 20 km.

The ahar-pyne system received a death-blow under the nineteenth-century British colonial regime. The post-independent state was hardly better. In 1949, a Flood Advisory Committee investigating continuous floods in Bihar's Gaya district came to the conclusion that "the fundamental reason for recurrence of floods was the destruction of the old irrigational system in the district."

Of late, though, some villages in Bihar have taken up the initiative to re-build and re-use the system. One such village is Dihra.

It is a small village 28 km southwest of Patna city. In 1995, some village youths realised that they could impound the waters of the Pachuhuan (a seasonal stream passing through the village that falls into the nearby river Punpun) and use its bed as a reservoir to meet the village's irrigation needs. Essentially, this meant creating an ahar-pyne system

After many doubts, the village powers-that-be gave the go-ahead. Money was collected and work began in May 1995. After a month of shramdaan (voluntary labour) the villagers completed their work mid-June.

Their efforts have borne fruit. By 2000 AD, the ahar was irrigating 80 ha of land. The people grow two cereal crops and one crop of vegetables every year. The returns from the sale of what they produce are good. The village is no longer a poor one.

Bengal's Inundation Channel

Bengal once had an extraordinary system of inundation canals. Sir William Willcocks, a British irrigation expert who had also worked in Egypt and Iraq, claimed that inundation canals were in vogue in the region till about two centuries ago. Floodwater entered the fields through the inundation canals, carrying not only rich silt but also fish, which swam through these canals into the lakes and tanks to feed on the larva of mosquitoes. This helped to check malaria in this region. According to Willcocks, the ancient system of overflow irrigation had lasted for thousands of years. Unfortunately, during the Afghan-Maratha war in the 18th century and the subsequent British conquest of India, this irrigation system was neglected, and was never revived.

According to Willcocks, the distinguishing features of the irrigation system were:

1. the canals were broad and shallow, carrying the crest waters of the river floods, rich in fine clay and free from coarse sand;
2. the canals were long and continuous and fairly parallel to each other, and at the right distance from each other for purposes of irrigation;
3. irrigation was performed by cuts in the banks of the canals, which were closed when the flood was over.

Dungs or Jampois

Dungs or Jampois are small irrigation channels linking rice fields to streams in the Jalpaiguri district of West Bengal.

Cheruvu

Cheruvu are found in Chitoor and Cuddapah districts in Andhra Pradesh. They are reservoirs to store runoff. Cheruvu embankments are fitted with thoomu(sluices), alugu or marva or kalju (flood weir) and kalava (canal).

Kohli Tanks

The Kohlis, a small group of cultivators, built some 43,381 water tanks in the district of Bhandara, Maharashtra, some 250-300 years ago. These tanks constituted the backbone of irrigation in the area until the government took them over in the 1950s. It is still crucial for sugar and rice irrigation. The tanks were of all sizes, often with provisions to bring water literally to the doorstep of villagers.

Bhanadaras

These are check dams or diversion weirs built across rivers. A traditional system found in Maharashtra, their presence raises the water level of the rivers so that it begins to flow into channels. They are also used to impound water and form a large reservoir.

Where a bandhara was built across a small stream, the water supply would usually last for a few months after the rains.

They are built either by villagers or by private persons who received rent-free land in return for their public act.

Most Bandharas are defunct today. A very few are still in use.

Climate change and Rainwater Harvesting

Holocene (~ 11 kyr BP to present) climate fluctuations, including large spatial variation in Holocene monsoon and temperature over India are now well-resolved from various climate change proxies. A somewhat coherent, although unresolved, picture that emerges for India is as follows: as the aridity increased in the region as is evident from palaeoclimatic studies, people intensified rainwater harvesting as is seen from archaeological and historical evidences. Folk sayings such as ?capture rain where it falls? may have originated in response to the increased aridity in the Indian region over the last few millennia. Such climate fluctuations may have given rise to traditional village tanks, ponds and earthen embankments numbering more than 1.5 million that still harvest rainwater in 660,000 villages in India and encourage growth of vegetation in commons and agro-ecosystems. Indeed, a specific emphasis through the long sweep of history on management of rainwater harvesting systems in ancient texts, such as Rigveda

Addressing water problem holds the promise in future for a world compounded by climate change, growing population and decreasing water-impounding area of traditional tanks due to urbanization. We will have to take into account the large-scale, natural climate variations as well as human-induced climate change in the management of natural, social and economic systems. Alternative to ecologically damaging, socially intrusive and capital-intensive water management projects that fail to deliver their desired benefits, it would be useful investing in decentralized facilities, efficient technologies and policies, and human capital to improve overall productivity rather than to find new sources of water supply. Such efforts would need to be encouraged with innovative policy regimes that concurrently promote rainwater harvesting.

INDIAN MONSOON

Monsoon is a season looked forward with great anticipation all over India. Cool winds begin to blow from the southwest often with gale-like intensity, followed soon after by welcome showers of copious rain. Skies remain cloudy for nearly three months, the whole scene of India gets transformed and tremendous activity is witnessed all over the country. Sometimes, the starting of the southwesterly winds may get delayed, the quantum of water received may get reduced and there may be long breaks in the monsoon current causing anxiety among the people wholly dependent on this heaven-born resource. Monsoon, therefore, is of vital concern to all of us.

During the summer months (June-September) cool winds charged with moisture blow from the SW. Due to the existence of the high Sahya?dri barrier across their flow path, the low-lying moisture-rich currents are forced to rise, become cooled, condense and precipitate in the form of rain on the windward slopes of the Sahya?dri. The Sahyadri has played an important role in fashioning the climate of India and the timing of elevation of this mountain range is therefore crucial to our understanding the periodicity and predictability of the Indian monsoon. Similarly, the winds blowing northward over the Bay of Bengal are obstructed and diverted westwards by the Himalaya causing copious amounts of rainfall all along the foothill region and the Gangetic valley in front of the mountains. These features point to the important role played by oceans, and the major mountain ranges of India the Sahyadri and the Himalaya in the distribution of rainfall which is characteristically Indian and termed therefore the Indian summer monsoon. (Geological Society of India, Selected Editorials)

Water Harvesting in Ancient India

Traditionally, such systems have been integrated with agro-forestry and ethno-forestry practices, and remain useful in contemporary conservation and ecological restoration of degraded ecosystems. A systematic support to local innovations on rainwater harvesting could provide substantial amounts of water. Rural and urban water use, restoration of streams for recreation, freshwater fisheries, and protection of natural ecosystems are all competing for water resources earlier dedicated to food production. Under such circumstances, decentralized rainwater harvesting adaptations prove efficient. Traditional systems would become more efficient if scientific attempts are combined to enhance the productivity of local knowledge. With an insightful policy, rainwater harvesting can be promoted as a core adaptation strategy for achieving the global security and sustainability of water resources in an era of anthropogenic climate change. Rainwater harvesting in response to climate extremes enhances the resilience of human society. In a world confronting local and global changes, building resilience of human society to absorb shock, learn and develop, would depend on sound knowledge of the historical adaptive processes that are still functional. An integrated perspective of traditional knowledge on adaptation strategies, such as the rainwater harvesting system, is particularly useful to comprehend vulnerability and adaptation to environmental stresses at the local scale. I believe that neither the water policy nor the climate policy discussions seem to notice the worth of rainwater harvesting as an adaptation to climate change, especially in urban areas where water resources are fast depleting due to rapid increase in population and unrestricted use of water. Studies of historical societal adaptations to climate fluctuations provide insights into possible responses of modern societies to future climate change and sustainable management of water resources

Flooding, a common feature of Urban calender

The core of the present SWD system in city (Mumbai) is about 70 year old, comprising of about 400 km of underground drains and laterals built on the basis of population and weather conditions. The old SWD system is capable of handling rain intensity of 25 mm per hour at low tide. If the rain intensity is more than 25 mm per hour and high tide occurs, there is always a possibility of water logging.

Since the discharge of all the storm water and treated sewage is into the Arabian Sea, tidal variation has a major bearing in the system of storm water drainage (SWD) resulting in flooding and water logging during heavy rains and recession of water during low tide.

Rainwater Harvesting India

Why discuss Rainwater Harvesting in India?

In India, rainwater harvesting is an ancient tradition. From as far back as 4500 BC, the simplest of earthworks in Thar Desert and Rajasthan, would harvest water from the falling rain. These simplest forms of rainwater harvesting would evolve in accordance to the eco-regions within India's borders. Using rivers, floods, monsoon, underground rivers, surface water and the earth itself, the ancient cast of pallar (water managers) have been respected for thousands of years. Rainwater harvesting in India is more than an age old tradition that varies from region to region, rainwater harvesting is an integral part of Indian identity and cultural history, that without, India would never have been.

Vedic culture did not create rainwater harvesting as it was already being done (although rudimentary still quite effective) in the Thar and Rajasthan deserts long before the Harappan civilization in 2600 BC. Monsoon has a way of dictating how life in any given region will profit from determined amounts of rainfall. Once construction techniques began to improve, more elaborate structures could be undertaken to access water management issues in creative and innovative ways. Not only household water needs, but also farm and irrigation in even the most remote places such as the Thar Desert. To this very day, wells known as "kuis" or "beris" that collect fallen rainwater and prevent evaporation still exist, were first built by caravan travelers that had determined routes through the desert.

More developed wells called "kundis" or "kunds" are still used for drinking water, while "bundela" and "chandela" tanks with steps leading down into them were surrounded by pavilions, gardens, orchards to glorify the king. The type of rainwater harvesting techniques used in India, vary in accordance to the region. In northern India the Himalayan regions use glacier water and artificial glaciers to have water year round. The western Himalayan region which gives life to the Ganga also uses similar ditch technologies to replenish the underground flow of water and produce massive agriculture in an otherwise desolate area. The Thar Desert is an exceptional example of ingenuity and sustainability in low-tech communities. But ingenuity and creativity are what mark most about any region in India, as the solutions have time and time again come from the climatic and geographical conditions in which peoples had found themselves and still find themselves to this very day.

Whether they are harvesting rain from their rooftops, or courtyards, open community lands from artificial wells, monsoon run-off from the water of swollen streams or stored in various bodies or even harvest water from flooded rivers. Rainwater harvesting managers, called “pallar” are an officially recognized cast in India that deserves respect and honor. Usually “pallar” inherit their skills, and perform their service usually in accordance to region. The pallar are not trained in great universities from around the world, much less any inside of India, the pallar learn their abilities from one generation to the next and the most important part of engineering itself, experience. Years of careful observation on a day to day basis, when water needs effect everyday life, and even survival, their creative minds invent solutions that bridge the frontier of time and technology. Ingenuity and creativity in such largely diverse scales are responsible for the plethora of innovative ideas that come from this humble Indian cast.

Prehistoric India brought rainwater harvesting solutions as modern day India also does. These solutions which are diverse and innovative bring new insight into the world of rainwater harvest the world over. Insights that should be studied, and understood, not merely as a science but as cultural identity and a way of thinking thatâ€™s roots can be traced to antiquity. As India has so many different regions, it also confronts many different solutions for such a basic and essential human need as a single drop of water. Flood water, post-monsoon drought, underground river collectors, surface water aqueducts, and even evaporation proof community wells for drinking as well as irrigation and other methods of rainwater harvesting; the ancient art of Indian pallar is a tradition that should be respected and understood by anyone interested in better and more ecological ways to use the sky-gift of natural rain.

8

USE OF MODERN TRACTORS

A tractor is a vehicle specifically designed to deliver a high tractive effort (or torque) at slow speeds, for the purposes of hauling a trailer or machinery used in agriculture or construction. Most commonly, the term is used to describe a farm vehicle that provides the power and traction to mechanize agricultural tasks, especially (and originally) tillage but nowadays a great variety of tasks. Agricultural implements may be towed behind or mounted on the tractor, and the tractor may also provide a source of power if the implement is mechanised.

The word tractor was taken from Latin, being the agent noun oftrahere "to pull". The first recorded use of the word meaning "an engine or vehicle for pulling wagons or ploughs" occurred in 1901, displacing the earlier term traction engine

HISTORY

The first powered farm implements in the early 19th century were portable engines - steam engines on wheels that could be used to drive mechanical farm machinery by way of a flexible belt. Around 1850, the first traction engineswere developed from these, and were widely adopted for agricultural use. The first tractors were steam-powered plowing engines. They were used in pairs, placed on either side of a field to haul a plow back and forth between them using a wire cable. Where soil conditions permitted (as in the United States) steam tractors were used to direct-haul plows, but in the UK and elsewhere plowing engines were used for cable-hauled plowing instead. Steam-powered agricultural engines remained in use well into the 20th century until reliable internal combustion engines had been developed.

In 1892, John Froelich invented and built the first gasoline/petrol-powered tractor in Clayton County, Iowa, USA. After receiving a patent Froelich started up the Waterloo Gasoline Engine Company, investing all of his assets which by 1895, all would be lost and his business resigned to become a failure.

Charles W. Hart and Charles H. Parr developed a two-cylinder gasoline engine and set up their business in Charles City, Iowa. In 1903 the firm built fifteen "tractors". A term with Latin roots coined by Hart and Parr and a combination of the words traction and power. The 14,000 pound #3 is the oldest surviving internal combustion engine tractor in the United States and is on display at the Smithsonian National Museum of American History in Washington D.C. The two-cylinder engine has a unique hit-and-miss firing cycle that produced 30 horsepower at the belt and 18 at the drawbar.

In Britain, the first recorded tractor sale was the oil-burning Hornsby-Ackroyd Patent Safety Oil Traction engine, in 1897. However, the first commercially successful design was Dan Albone's three-wheel Ivel tractor of 1902. In 1908, the Saunderson Tractor and Implement Co. of Bedford introduced a four-wheel design, and went on to become the largest tractor manufacturer outside the U.S. at that time.

While unpopular at first, these gasoline-powered machines began to catch on in the 1910s when they became smaller and more affordable. Henry Ford introduced the Fordson, the first mass-produced tractor in 1917. They were built in the U.S., Ireland, England and Russia and by 1923, Fordson had 77% of the U.S. market. The Fordson dispensed with a frame, using the strength of the engine block to hold the machine together. By the 1920s, tractors with a gasoline-powered internal combustion engine had become the norm.

Safety

Agriculture in the United States is one of the most hazardous industries, only surpassed by mining and construction. No other farm machine is so identified with the hazards of production agriculture as the tractor. Tractor-related injuries account for approximately 32% of the fatalities and 6% of the non-fatal injuries in agriculture. Over 50% is attributed to tractor overturns.

The roll over protection structure (ROPS) and seat belt, when worn, are the two most important safety devices to protect operators from death during tractor overturns.

Modern tractors have rollover protection systems (ROPS) to prevent an operator from being crushed if the tractor overturns. It is important to remember that the ROPS does not prevent tractor overturns. Rather, it prevents the operator from being crushed during an overturn. This is especially important in open-air tractors, where the ROPS is a steel beam that extends above the operator's seat. For tractors with operator cabs, the ROPS is part of the frame of the cab. A ROPS with enclosed cab further reduces the likelihood of serious injury because the operator is protected by the sides and windows of the cab.

ROPS were first required by legislation in Sweden in 1959. Before ROPS were required, some farmers died when their tractors rolled on top of them. Row-crop tractors, before ROPS, were particularly dangerous because of their 'tricycle' design with the two front wheels spaced close together and angled inward toward the ground. Some farmers were killed by rollovers while operating tractors along steep slopes. Others have been killed while attempting to tow or pull an excessive load from above axle height, or when cold weather caused the tires to freeze to the ground, in both cases causing the tractor to pivot around the rear axle.

For the ROPS to work as designed, the operator must stay within the protective frame of the ROPS. This means the operator must wear the seat belt. Not wearing the seat belt may defeat the primary purpose of the ROPS.

Applications and variations

Farm tractor applications

The most common use of the term "tractor" is for the vehicles used onfarms. The farm tractor is used for pulling or pushing agricultural machinery or trailers, for plowing, tilling, disking, harrowing, planting, and similar tasks.

A variety of specialty farm tractors have been developed for particular uses. These include "row crop" tractors with adjustable tread width to allow the tractor to pass down rows of corn, tomatoes or other crops without crushing the plants, "wheatland" or "standard" tractors with non-adjustable fixed wheels and a lower center of gravity for plowing and other heavy field work for broadcast crops, and "high crop" tractors with adjustable tread and increased ground clearance, often used in the cultivation of cotton and other high-growing row crop plant operations, and "utility tractors", typically smaller tractors with a low center of gravity and short turning radius, used for general purposes around the farmstead. Many utility tractors are used for non-farm grading, landscape maintenance and excavation purposes, particularly with loaders, backhoes, pallet forks and similar devices. Small garden or lawn tractors designed for suburban and semi-rural gardening and landscape maintenance also exist in a variety of configurations.

Some farm-type tractors are found elsewhere than on farms: with large universities' gardening departments, in public parks, or for highway workman use with blowtorch cylinders strapped to its sides and a pneumatic drill air compressor permanently fastened over its power take-off. These are often fitted with grass (turf) tyres which are less damaging to soft surfaces than agricultural tires.

ENGINEERING TRACTORS

The durability and engine power of tractors made them very suitable for engineering tasks. Tractors can be fitted with engineering tools such as dozer blade, bucket, hoe, ripper, and so on. The most common attachments for the front of a tractor are dozer blade or a bucket. When attached with engineering tools the tractor is called an engineering vehicle.

A bulldozer is a track-type tractor attached with blade in the front and a rope-winch behind. Bulldozers are very powerful tractors and have excellent ground-hold, as their main tasks are to push or drag things.

Bulldozers have been further modified over time to evolve into new machines which are capable of working in ways that the original bulldozer can not. One example is thatloader tractors were created by removing the blade and substituting a large volume bucket and hydraulic arms which can raise and lower the bucket, thus making it useful for scooping up earth, rock and similar loose material to load it into trucks.

A front-loader or loader is a tractor with an engineering tool which consists of two hydraulic powered arms on either side of the front engine compartment and a tilting implement. This is usually a wide open box called abucket but other common attachments are a pallet fork and a bale grappler.

Other modifications to the original bulldozer include making the machine smaller to let it operate in small work areas where movement is limited. There are also tiny wheeled loaders, officially called Skid-steer loaders but nicknamed "Bobcat" after the original manufacturer, which are particularly suited for small excavation projects in confined areas.

The word 'tractor' originated from the Latin word 'trahere', meaning 'pull'. It is an engine or vehicle for pulling wagons or plows. Today, tractors are used for drawing in, towing or pulling objects that are extremely hard to move. One commonly sees a tractor on farms used to push agricultural machinery or trailers that plow or harrow fields. The first tractors from the 1800s and early 1900s, were powered by steam engines. These tractors were phased out due to the instability of the steam engine that caused explosions, or trapped the driver in a belt driven attachment. The successors were built with an internal combustion engine. Technology has found its way into modernizing the tractor. Nowadays, one finds Global Positioning Systems (GPS) and on-board computers attached to farm tractors. With the use of advance technology, corporate-scale farms are using a combination of unmanned tractors and manually driven vehicles to plow the fields.

A tractor is a vehicle specifically designed to deliver a high tractive effort at slow speeds, for the purposes of hauling a trailer or machinery used in agriculture or construction. Most commonly, the term is used to describe the distinctive farm vehicle. Agricultural implements may be towed behind or mounted on the tractor, and the tractor may also provide a source of power if the implement is mechanized. Another common use of the term is for the power unit of a semi-trailer truck. The first tractors, powered by steam engines, were phased out followed by the internal combustion engine.

Modern tractors are built with a rollover protection system (ROPS) that protects the operator from being crushed in the event that the vehicle rolls over. It was the New Zealand legislation that first required the ROPS to be built in to tractors in the 1960s. Before this system was required, many farmers were killed in accidents when the tractors rolled on top of them, crushing them in the process. Usually these accidents happen when tractors were driven along steep

History of Tractors

In 1892, John Froelich built the first practical gasoline-powered tractor in Clayton County, Iowa. Only two were sold, and it was not until 1911, when the Twin City Traction Engine Company developed the design, that it became successful. The classic farm tractor is a simple open vehicle, with two very large driving wheels on an axle below and slightly behind a single seat (the seat and steering wheel consequently are in the center), and the engine in front of the driver, with two steerable wheels below the engine compartment.

This basic design has remained unchanged for a number of years, but enclosed cabs are fitted on almost all modern models, for reasons of operator safety and comfort. Originally, plows and other equipment were connected via a draw-bar, or a proprietary connecting system; prior to Harry Ferguson patenting the three-point hitch. Recently, Bobcat's patent on its front loader connection has expired; and compact tractors are now being outfitted with quick-connect attachments for their front-end loaders.

There are also lawn tractors. Cub Cadet, Husqvarna, John Deere, Massey Ferguson and Toro are some of the better-known brands

Uses of Tractors

Farm implements can be attached to the rear of the tractor by either a drawbar or a three-point hitch. The three-point hitch was invented by Harry Ferguson and has been standard since the 1960s. Equipment attached to the three-point hitch can be raised or lowered hydraulically with a control lever. The equipment attached to the three-point hitch is usually completely supported by the tractor. Another way to attach an implement is via a Quick Hitch, which is attached to the three-point hitch. This enables a single person to attach an implement quicker and put the person in less danger when attaching the implement.

In some farm-type tractor, blowtorch cylinders strapped to its sides are used with a pneumatic drill air compressor permanently fastened over its power take-off. These are often fitted with grass (turf) tyres which are less damaging to soft surfaces than agricultural tires. Supposedly industrial bar tires are less damaging to lawns and soft surfaces than agricultural tires, but provide similar traction, and have the benefit of being self-cleaning. Often, these can be seen on road construction backhoes. Equipment costs account for about 20 percent of the total costs of production. The cost for depreciation, interest and insurance can range from $25 to $60 per acre depending on the crop and how intensively the equipment is used.

Operation

Modern farm tractors usually have five foot-pedals for the operator on the floor of the tractor. The pedal on the left is the clutch. The operator presses on this pedal to disengage the transmission for either shifting gears or stopping the tractor. Two of the pedals on the right are the brakes. The left brake pedal stops the left rear wheel and the right brake pedal does the same with the right side. This independent left and right wheel braking augments the steering of the tractor when only the two rear wheels are driven. This is usually done when it is necessary to make a tight turn. The split brake pedal is also used in mud or soft dirt to control a tire that spins due to loss of traction. The operator presses both pedals together to stop the tractor. For tractors with additional front-wheel drive, this operation often engages the 4-wheel locking differential to help stop the tractor when travlling at road speeds.

A fifth pedal just in front of the seat operates the rear differential lock (diff. lock) which prevents wheelslip. The differential lock allows the outside wheel to travel faster than the inside one during a turn. However, in traction conditions on a soft surface, the same mechanism could allow one wheel to slip, thus preventing traction to the other wheel. The diff. lock overrides this, causing both wheels to supply equal traction. Care must be taken to unlock the differential, usually by hitting the pedal a second time, before turning, since the tractor cannot perform a turn with the diff. lock engaged.

Useful Steps in Operation of Tractors

- The tractor should be proceeded slowly and carefully
- Ditches, gullies and steep embankments should be avoided as far as possible
- Be aware that a towed trailer or heavy implement can easily upset a tractor's balance and it can overturn the tractor causing hazards
- Accelerate gradually against gravity's resistance. Tractors tend to flip over backwards while going uphill, especially with a rear implement or when towing

- Small rocks, holes and bumps should be covered carefully. They can quickly disturb a tractor's equilibrium
- If possible, always travel directly up or down a hill, never across the slope
- If cross slope operation is necessary, always turn down the hill with the slope, never uphill into the slope
- While transporting a load uphill, always attach the load to the front of the tractor and back up the hill
- When towing a load downhill, use a low gear and let the tractor's engine act as a brake.

Types of Tractors

Compact Utility Tractor

As the name suggests these tractors are modified and assembled to serve various purposes. CUT, as it is commonly called is a smaller version of an agricultural tractor but designed primarily for landscaping and estate management type tasks rather than for planting and harvesting on a commercial scale. Typical CUTs range in from 20 to 50 horsepower (15-37 kW) with available power take off (PTO) horsepower ranging from 15 to 45 hp (11-34 kW). CUTs are often equipped with both a mid-mounted PTO and a standard rear PTO, especially those below 40 horsepower (30 kW). The mid-mount PTO shaft typically rotates at/near 2000 rpms and is typically used to power such implements as mid-mount finish mower, a front mounted snow blower or front mounted rotary broom. The rear PTO is standardized at 540 rpms for the North American markets, but in some parts of the world, a dual 540/1000 rpm PTO is standard and implements are available for either standard in those markets. One of the most common attachment for a Compact Utility Tractor is the front end loader or FEL. All modern CUTs feature a government mandated roll over protection structure (ROPS) just like agricultural tractors. Although less common, compact backhoes are often attached to compact utility tractors.

Farm Tractor

The most common use of the term tractor is for the vehicles used on farms. The farm tractor is used for pulling agricultural machinery or trailers, for plowing, harrowing and similar tasks. The classic farm tractor is a simple open vehicle with two very large driving wheels on an axle below and slightly behind a single seat (the seat and steering wheel consequently are in the center) and the engine in front of the driver with two steerable wheels below the engine compartment. This basic design has remained unchanged for a number of years, but now enclosed cabs are available for many models of farm tractor. There are usually four foot-pedals, for the operator, on the floor of a tractor. The pedal on the left is the clutch. The operator presses on this pedal to disengage the transmission for either shifting gears or stopping the tractor. Two of the pedals on the right are the brakes. The left brake pedal stops the left rear wheel and the right brake pedal does the same with the right side. This independent left and right wheel braking augments the steering of the tractor when only the two rear wheels are driven. This is usually done when it is necessary to make a tight turn.

Engineering Tractor

These tractors are fitted with engineering tools such as bucket, hoe, ripper, dozer blade, and so on. The most common attachments for the front of a tractor are the dozer blade or a bucket. Attached with engineering tools, the tractor is called an engineering vehicle and it is capable of performing complex processes. A bulldozer is one of this kinds of tracked-type tractor attached with blade in the front and a rope-winch behind. Bulldozers are very powerful tractors and have excellent ground-hold, as their main tasks are to push or drag things. Bulldozers have been further modified over time to evolve into new machines which are capable of working in ways that the original bulldozer cannot. One example is that loader tractors are created by removing the blade and substituting a large volume bucket and hydraulic arms which can raise and lower the bucket, thus making it useful for scooping up earth, rock and similar loose material to load it into trucks.

Front Loader

A front-loader or loader is a tractor with an engineering tool which consists of two hydraulic powered arms on either side of the front engine compartment and a tilting implement. This is usually a wide open box called a bucket but other common attachments are a pallet fork and a bale grappler. Other modifications to the original bulldozer include making the machine smaller to let it operate in small work areas where movement is limited. There are also tiny wheeled loaders, officially called skid-steer loaders but nicknamed "Bobcat" after the original manufacturer. These types of tractors are particularly suited for small excavation projects in confined areas.

Backhoe Loader

The Backhoe Loader is a special purpose tractor used mainly in the construction industry. As the name implies, it has a loader assembly on the front and a backhoe on the back. When both the loader and the backhoe are permanently attached, it is almost a huge crane. When the backhoe is permanently attached, the machine usually has a seat that can swivel to the rear to face the hoe controls. Removable backhoe attachments almost always have a separate seat on the attachment. Backhoe-loaders are very common and can be used for a wide variety of tasks like construction, small demolitions, light transportation of building materials, powering building equipment, digging holes, breaking asphalt and paving roads. Some buckets have a retractable bottom, enabling them to empty their load more quickly and efficiently. Buckets with retractable bottoms are also often used for grading and scratching off sand. The front assembly may be a removable attachment or permanently mounted. Often the bucket can be replaced with other devices or tools.

Garden Tractor

These tractors are small in size and used mainly for domestic plowing and gardening. Garden Tractors have stronger frames, axles and transmissions. Garden Tractors are generally capable of mounting other implements such as harrows, cultivators, sweepers, rollers and dozer-blades. Garden Tractors (also called Mini Tractors) are small, light and simple tractors designed for use in domestic gardens. These are usually designed primarily for cutting grass, being fitted with horizontal rotary cutting decks like ride-on mowers. These type of tractors generally have a horizontally-mounted engine with a belt-drive to a transaxle-type transmission (usually of 4- or 5-speeds, although some my also have two-speed reduction gearboxes or hydraulic gearboxes).

EPA Tractor

An EPA tractor is made just like an automobile, truck or lorry, without the passenger space behind the front seats, equipped with two gearboxes in a row to give the speed of loading heavy weights..

Road Tractor

They are heavy-duty vehicles with large engines and several axles. The majority of these tractors are designed to pull long semi-trailers, most often to transport freight of some kind over a significant distance, and is connected to the trailer with a fifth wheel coupling.

Locomotive Tractor

Amalgamating machines, electrical generators, controls and devices that comprise the traction component of railway vehicles, these tractors are built to move heavy vehicles. They are used mainly for rail car movers.

Tractor Hazards and Safety

Deaths involving tractors contribute to agriculture having one of the highest on-the-job annual death rates -- 20.3 deaths for 100,000 workers nationally in the United States of America. In a 1999 study conducted by the Deep-South Center for Agricultural Health and Safety, it was found that, on average, 21 adults died each year in Florida agriculture in recent years. At least 74 of those deaths involved machinery, including tractors overturning and crushing their drivers. Across the nation, approximately 36 per cent of the agricultural work fatalities in 1997 involved a tractor, and 19 percent involved other machinery, according to the National Safety Council.

Safer tractor operations and the use of protective equipment could prevent many of the deaths and injuries involving tractors. Owners, managers, and tractor operators all play a major role in making tractor operations safer. Owners have responsibilities for providing a safer environment and safer equipment. Managers coordinate maintenance designed to ensure the safety of equipment, as well as safety policies and programs. The final responsibility for making the operation of tractors safer lies with the tractor operators themselves. It's often only when an incident occurs that people become focused on what might have prevented injury or loss of life. The following pages give guidance to agricultural employees on making tractor use safer.

Implementation of Safety Rules

- Dismounting load from a tractor when the engine is running is not allowed, unless the operation requires it
- Consumption of alcohol or drugs, including medications that may impair judgment, during work hours or for four hours before starting work should be prohibited
- Always use hearing protection
- Carbon monoxide hazards can be detected by signs of severe headeache. Check to see if the exhaust is in a location that exposes the operator to exhaust fumes. There have been instances of carbon monoxide exposure using an open-station tractor in the field
- If the tractor has a ROPS, securely fasten the seat belt. Do not use seat belts on tractors that don't have ROPS
- Where possible, avoid operating the tractor near ditches, embankments, and holes
- Reduce speed when turning and crossing slopes and on rough, slick, or muddy surfaces

- Stay off slopes that are too steep for safe operation
- Watch carefully for obstacles and other hazards in the tractor's path, especially at the end of rows, on roads, around trees and in other blind corners.

Tractor Sales and Farmers

Facing market saturation in the traditional markets of the north west (Punjab, Haryana, eastern Uttar Pradesh) tractors sales began a slow and slight decline. By 2002 sales went below 200,000. Manufacturers scrambled to push into eastern and southern India markets in an attempt to reverse the decline, and began exploring the potential for overseas markets. Sales remained in a slump, and added to the market saturation problems also came increased problems of "prestige" loan defaults, where farmers who were not financially able took tractors in moves to increase their families prestige. There are also reported increased misuse of these loans for buying either lifestyle goods, or for social functions. Government and private banks have both tightened their lending for this sector adding to the industry and farmers woes.

By 2004, a slight up tick in sales was witnessed once again. But by 2006 sales once again were down to 216,000 in 2007-08 have slid further to just over 200,000.

The Road Ahead

There has been recent trends in the development of tractors. Two new forms of modernized tractor is a matter of discussion and research among agricultural scientists all over the world, one a 'farmer's vehicle', a universal agricultural machine which could complete all operations in a single pass. The other is a single purpose 'tractor robot', an operator-free agricultural vehicle. The pros and cons of each option are enumerated and requirements for the development of a tractor-robot are in discussion. The tractor-robot has a system of cameras for a round 'vision', a satellite positioning system, movement sensors, control computer and automated motor controls. A prototype of a tractor-robot is presented and assessed; it is equipped with a self-locking cab which can be opened manually if needed, and a standard three-point PTO (power take off) for implements. Tractors are an indispensable agricultural machinery which is all set to bring about revolutionary changes in the world agricultural production.

The tractor as a modern agricultural tool

The modern tractor has evolved over the years. Powered farm implements in the nineteenth century where steam engines. These were used to drive farm machinery mechanically. This was the birth of modern day tractors and was used for agricultural purposes. This was later followed by internal combustion gasoline or petrol powered tractors. At first the gasoline powered tractors were unpopular until the Fordson brand, powered by internal combustion engine was introduced and became a hit.

Modern tractors are classified into two wheel drive and four wheel drives. Most of the tractors in the market are classical open vehicles with two huge rear wheels and small front wheels which are very key to steering its direction. The engine is strategically placed for high traction and balance. Caterpillars and crawlers became popular in the early thirties in heavy or wet soils due to their superior floatation and traction.

The basic use of a tractor is in the agricultural field though innovative manufacturers have come up with specific tractors for specific. The row crop tractor is specifically for tending crops from the time they sprout to harvest time and combines well with the cultivator. We also have the garden or mini tractors which are very light and used in small scale or personal gardens. Two wheel tractors are more or less like the garden counterparts and have a huge loyalty in the United States of America and beyond. Lastly and not the least we have the orchard tractors which are tailored to pass under orchard trees unperturbed.

All these innovations and tractor designed are geared towards precision agriculture and popularised by the modern lifestyles of reduction of manual labour by employing use of machines. All in all, it should be noted that use of machinery is not a bed of roses if safety is not checked properly. In the United States of America, agriculture is one of the most hazardous industries behind mining and construction. When looking for tractors to buy, the safety features are some of the aspects that will lead you to decide the kind of tractor fit for you.

The country inherited a stagnant agriculture at the time of Independence. The traditional tools and implements relied mostly on human and animal power and used a negligible amount of commercial energy. However, successive governments realized the importance of agriculture and initiatives were taken for the growth of this sector. Increased investment in irrigation infrastructure, expansion of credit, marketing, and processing facilities.

Till the 1950s, use of tractors for agriculture was very limited. Tractor manufacturing in India started in 1961 with aggregate capacity to manufacture 11 000 tractors. Joint efforts made by the government and private sector have led to steady increase in the level of mechanization over the years.

Given that rains are not always timely and evenly distributed, farmers prefer pump sets as a more reliable and assured source of irrigation; as a result, energization of pump sets have been increasing rapidly. As on 31 March 2004, 14.1 million pump sets had been energized. Maharashtra has the maximum number of energized pump sets (2.4 million), followed by Andhra Pradesh (2.3 million). Earlier, the average capacity of the pump sets was 3.68 kW and a pump set on an average consumed

6004 kWh of electricity in that year (Central Electricity Authority,2005). However, owing to insufficient electricity supplies, some farmers have also procured diesel pump sets as a standby. In the recent past, concerted efforts of the government has led to an introduction of biomass and solar photovoltaic based pumping systems

As a result of increased mechanization in agriculture, crop production and rural agro processing emerged as one of the major consumers of commercial energy. The share of mechanical and electrical power in agriculture increased from 40% in 1971/72 to 84% in 2003/04. The availability of farm power per unit area (kW/ha) has been considered as one of the parameters of expressing the level of mechanization. Power availability for carrying out various agricultural operations has increased from 0.3 kW/ha in 1971/72 to the tune of 1.4 kW/ha in 2003/04.

Connected load in the agriculture sector in 2004 was estimated to be 51.84 GW, the number of consumers being 12.8 million. The electricity consumption in agriculture during 2003/04 was 87 089 GWh (second highest)?24.13% of the total electricity consumption. There was an increase of 3.08% in the electricity sales to the agriculture sector in 2003/04 over 2002/03 (CEA 2005). Electricity consumption in agriculture sector has been increasing mainly because of greater irrigation demand for new crop varieties and subsidized electricity to this sector. Moreover, due importance is not given to proper selection, installation, operation, and maintenance of pumping sets, as a result of which they do not operate at the desired level of efficiency, leading to huge waste of energy.

Agriculture (plantation/food) consumed 7 123 thousand tonnes of HSD (high-speed diesel) in 2003/04, accounting for 19.2% of the total HSD consumption during the year. Consumption of LDO (light diesel oil) and furnace oil for plantation in 2003/04 was 44 000 and 243 000 tonnes, respectively, accounting for 2.7% of the total LDO and 2.9% of the total furnace oil consumed in the country. Consumption of furnace oil for transport (agriculture retail trade) in the agriculture sector was 94 thousand tonnes (Ministry of Power and Natural Gas 2004).

What are tractors for?

As promised, here is the thrilling second installment of our multi-part series, "What are tractors for?"

Last week, Siri wrote about what is surely the most common use of tractors in agriculture - turning a field with grass or other plants growing on it into a field of bare dirt. In most of today's agriculture, this means huge tractors pulling huge implements. On small farms like ours, we use many of the same implements to do this job - plows, chisel plows, discs - just smaller versions of them.

Once a field is prepared for planting, on our farm the next step is putting seeds or seedlings in the ground. On bigger farms than ours, farmers may plant only one or two crops once or twice a year. Market farms like ours, though, need to have a consistent supply of vegetables for months on end. You can't achieve this by planting acres of lettuce and kale just once in the spring. Instead, we plant smaller quantities, just a bed or two at a time, of a diverse mix of vegetables every week. This is where our smaller tractors, and especially our new 1950 Farmall Cub come into play. Our Cub does four jobs on the farm that modern tractors aren't as good at doing. It creates our beds, it marks rows for transplanting, it seeds, and it cultivates.

We make our beds on the farm by driving the Cub over recently tilled soil in as-straight-as-possible lines back and forth every 5 feet. Every bed on our farm is 5 feet wide by 125 feet long. Coincidentally, the Cub has 8 inch wide tires that are spaced 5 feet apart, center to center. This means, when we drive the tractor over freshly tilled soil, the tires create a depression where we can walk and leave a 4 foot wide raised bed between the tires for us to plant into. On our farm we set up flags every 5 feet on both ends of the bed for the tractor driver to aim at, ensuring that our beds are mostly straight and well spaced.

At the same time that the Cub is marking a bed, it might also be marking rows for transplanting. For ease of cultivation (which I'll discuss shortly), all the rows of vegetables within every bed are spaced identically. That is to say, we plant a maximum of three rows per bed, each spaced 15 inches apart. The Cub marks rows to transplant into by setting wheels down on the bed-top that make a line in the soil the length of the bed. Once a bed is marked, we'll return with trays of seedlings to transplant into one, two, or all three rows depending on the plants' need for space.

The other way we get rows of vegetables is by using our tractor mounted seeder. This year we purchased three Jang Seeders, which are ganged together and mounted on the back of the Cub. By lowering the rear hydraulics on the tractor, the seeders are set on the ground. As the tractor drives forward, the ground turns the front wheel of the seeder. This wheel, in turn, turns a barrel inside the seed hopper which picks up one seed at a time and drops it into the ground. The seeder then covers the seeds and compresses the soil behind them.

The last thing the Cub does is cultivate. The word cultivate is a funny one, because it has two agricultural meanings. The meanings most people know are "to grow plants," or "to prepare land for farming." That is to say, all the plowing, discing, and tilling Siri talked about in the prior post. What we mean when we say cultivate is to kill weeds. That is to say, hand weeding, hoeing, and tractor cultivating with the Cub. Beneath the belly of the Cub are mounted a set of knives that can be raised and lowered with the tractor's hydraulics.

To cultivate with the tractor, we move these knives so that they disturb the ground in the space between the rows of vegetables, killing only the weeds. Tractor cultivation is very tricky, for if the tractor driver goes off course even a little it can mean killing lots of vegetables. The engine and seat of the cub are offset to the left so the driver can keep a close watch on the knives and the vegetables.

Most of today's farms use chemicals to kill weeds, and no company that I know of has manufactured a cultivating tractor since the 1980s. So on farms like ours, which don't use chemicals for weed control, you'll often find old fashioned tractors like our Farmall Cub or an Allis Chalmers G for killing weeds.

There are tons of other uses for tractors on a farm. Today, I pulled an ancient cedar root out of the field. Next week we'll use the front end loader on a tractor like a cherry picker to help us work on the very high roof of our newest greenhouse. One of the funny things about becoming immersed in a subject is how you forget how little you knew before you began. After Siri wrote the first post about tractors, a number of people have admitted to us that they didn't actually understand much of the process necessary to turn a grassy field into a field of vegetables. Moreover, it seems that many of the details about what we do on the farm are lost on our audience because most people aren't familiar with what farmers actually do or the terms we use to describe our work. We ourselves didn't know much about the technicalities of growing food before we started farming. The role tractors play is but one small part of the whole process. But now, at least, you'll understand that part.

One of the most important technological changes in agriculture was the transition to the modern tractor.

The transition from horses and mules to modern farm machinery for tilling the soil and harvesting the crop did not begin until the 1920s.

Tractors are big machines. They can do many different kinds of work. Normally farmers use them to pull plows and to work other farm machines. Farm tractors are designed for use in flat fields.

A tractor is a machine that pulls or pushes a tool over land. Without tractor power and purpose-built implements, it's unlikely that anyone would look forward to work around the modern farm whether it is a small scale or lifestyle farm or a large commercial farm.

Modern farm tractors are effective and up-to-date agricultural power units. Changes have helped make them more efficient, safe, convenient, versatile and powerful.

Modern farm tractors may be divided into two general classes: the general purpose type; and the standard tread type.

The general purpose type of tractor, as its name implies, furnishes power for practically all farm work.

Not only does it perform all of the drawbar, belt and power shaft jobs, but, in addition puts speed and economy into the planting and cultivating of row crops.

New tractors can provide high levels of operator comfort, for example air-conditioning and radios in the cabin. These comfort features tend to lessen operator fatigue and therefore improve safety.

Modern farm tractors are equipped with receivers which using signals from Global Positioning System (GPS) satellites, allow precision hands-free plowing, planting and harvesting. Automated seed planters are synchronized with actual tractor speed by ground speed radar.

Modern tractors are equipped with roller protection system of ROPS, consisting mainly of a steel framework which if the operator is not thrown out, should prevent the weight fo the oveturned machine from crushing him.

Throughout the industrialized world, mechanization has become a cornerstone of production agriculture and the tractor is the modern symbol of farm mechanization.

9

TECHNOLOGIES IN BIO-RESOURCE ENGINEERING

Bioresource engineering is related to the applications of chemical engineering and agricultural engineering usually based on biological and/or agricultural feedstocks. Bioresource engineering is more general and encompasses a wider range of technologies and various elements such as biomass, biological waste treatment, bioenergy, biotransformations and bioresource systems analysis, and technologies associated with Thermochemical conversion technologies: combustion, pyrolysis, gasification, catalysis, etc. Biochemical conversion technologies: aerobic methods, anaerobic digestion, microbial growth processes, enzymatic methods, composting Products: fibre, fuels, feedstocks, fertilisers, building materials, polymers and other industrial products Management: modelling, systems analysis, decisions, support systems.

The impact of urbanization and increasing demand for land, food, and water presents engineers in a world with serious challenges. Little attention has been given to the interface between the biological world and traditional engineering in the past. It is the job of bioresource engineers to fill that gap. Agricultural and bioresource engineers develop efficient and environmentally-sensitive methods of producing food, fiber, timber, bio-based products and renewable energy sources for an ever-increasing world population. Some of the research in bioresource engineering include machine vision, vehicle modification, wastewater irrigation, irrigation water management, stormwater management, inside natural environment for animals and plants, sensors, non-point source pollution and animal manure management.

Accomplishments

A biosynthesis of silver nanoparticles (NPs) mediated by fungal proteins of Coriolus versicolor has been undertaken for the first time last year. Hydrogels have been used to separate As(V) from water. The importance of bioresource engineers are usually obscured, but it is they who are the scientists helping the environment. Most recently bioresource engineers were working on checking the thermal properties of compost from waste to use it as a renewable, low temperature heat source.

BIO-RESOURCE ENGINEERING PROJECTS

Bioresource engineering is a sub-field of chemical engineering. Engineers working on projects in this field attempt to change and manipulate the natural chemical processes of biological organisms. Bioresource engineering holds the promise of increasing food production and finding more environmentally sustainable methods of commercial production. It is a more recent form of chemistry, only beginning to find wide application during the early part of the 21st century. Its full potential has yet to be realized.

Agriculture

One form of bioengineering that has already found wide commercial use is in agriculture, where genetically altered plants, as well as different kinds of fertilizers, have proven useful for many purposes. As the world's population continues to grow, it is imperative that food stocks grow to match it. For this reason, bioengineering in agriculture has attracted great interest and investment from both private sources and public. State-of-the-art projects could greatly reduce hunger.

Materials

Numerous materials such as plastics typically involve production methods that are damaging to the environment by the pollution that they create and the resources that they use. One of the tasks of bioengineers is finding methods of production that use fewer nonrenewable resources and produce less pollution. One project that holds promise is to use the actual chemical processes of nature, such as when bacteria are used to produce certain chemical substances and materials.

Energy

Since the industrial revolution, most societies have produced their energy by burning fossil fuels. This has become a concern both for the pollution that such an energy process causes and also because of the fact that fossil fuels are nonrenewable resources. Many bioengineering projects are attempting to find other sources of fuel, such as that which can be produced with renewable plant life. Other projects focus cleaning up the pollution from energy production.

Waste

Waste disposal has always been a concern for human societies but has become even more of a concern as the planet has become more urbanized. Many projects in bioengineering are concerned today with finding more efficient ways of dealing with pollution cleanup and waste disposal. For instance, a potential use has been found for certain bacteria in cleaning up waste water. As the technology advances, it's likely that additional techniques will begin to find use.

Environmentalist

An environmentalist broadly supports the goals of the environmental movement, "a political and ethical movement that seeks to improve and protect the quality of the natural environment through changes to environmentally harmful human activities". An environmentalist is engaged in or believes in the philosophy of environmentalism.

Most politically inclined environmentalists identify themselves as greens and they have strong views on issues that concern the environment. The Green Parties are generally applied to those in the environmental movement working as volunteers, activists or paid staff. However, the term could also be applied to environmental scientists. Some environmentalists are also green anarchists, arguing that true environmental change cannot occur under the present economic model. Others are eco-socialists, who combine Marxism, socialism and ecology with environmentalism to produce an anti-capitalistideology that sees capitalism as the cause of environmental problems, social injustice and inequality.

Typically, environmentalists have conservationist views - in general, they advocate for the preservation, restoration, or enhancement of the natural environment.

Environmentalists are sometimes given names in an informal or derogatory context such as the terms "greenie" and "tree-hugger".

Strategic Themes in Agricultural and Bio-resource Engineering in the 21st Century

During the 20th century, agricultural engineering became a really multidisciplinary "eld of science. Besides traditional mechanical engineering, fundamental developments in microelectronics, systems and control engineering, chemical engineering and physics have opened wide perspectives for agricultural engineering. During the last century, agricultural engineering has elected major changes in agriculture with regard to labour productivity and specialization in the di!erent sectors of plant production and livestock production. Public concern for environmental issues, e.g. emissions of volatile compounds, water pollution through plant nutrients and agrochemicals, has resulted in research and technology development in the field of environmental engineering.

Hall (1992) described major trends initiated or influenced by the discipline of agricultural engineering. Early agricultural engineering activities were connected to horseshoeing, drainage, building construction, rural roads and steam power. Less support was given range problems. The main developments started when replacing animals by tractors and integration of units for sowing, cultivating and harvesting with the tractor. Electric power has improved productivity in livestock production and labour productivity on the farmsteads.

The possibilities of steel, fuel, lubricants and rubber played an important role in the design of the power units, the attached machinery and equipment. In the years before the Second World War (WWII), soil erosion and the use of water were important subjects for agricultural engineers. After WWII in Europe, the labour productivity in agriculture improved rapidly and the availability of industrial jobs, the adaptation of the farming system and the increasing demand for specific agricultural products supported further mechanization. The expertise of agricultural engineers on material characteristics, climate, control, structures, quality control of products was needed in these period. To benefit from the economy of scale and professional capabilities farmers specialized in arable crops (cereals, potatoes, sugar-beet), vegetable crops, dairy, poultry, pigs and greenhouse products. To-day, this specialization does raise questions with regard to sustainability and risk management. Farmers were interested in machinery devices and equipment not only for labour-saving or labour-serving reasons, but also for the quality of the process and the products. The productivity was also stimulated by scientific knowledge in other fields, such as: animal breeding, health and nutrition; plant breeding and nutrition; and the use of fertilizers and agrochemicals. Already in the period 1950-1970, application of chemicals resulted in requirements of the spraying devices with regard to distribution, accuracy and safety to the environment and the workers.

Inventions in the field of agricultural machinery, farm buildings and equipment influenced the production method to a larger extent, e.g. the cubicle, milking par-lour and milking tank in dairy production, the pick-up wagon for forage harvesting, precision drills and fertilizer broadcasters. From the 1950s onwards, other disciplines, such as electrical engineering, systems and control engineering, chemical engineering, and information technology, are crucial for the scientific progress in the enlarged field of agricultural engineering. This means that, on the one hand, agricultural engineering needs basic knowledge of these disciplines and at the other, that cooperation with these basic disciplines is necessary. This makes agricultural engineering a really multi-disciplinary working field.

Trends in the society

General

General trends in society are that the working life and the private situation will be mixed more and more, the workplace will be flexible, with more interest for human health and welfare and thus healthy food. Also new consumer demands require that fast changes, environment-friendly products and &green building and living' will be stimulated. Expansion of the use of Internet overs new possibilities worldwide for the distribution of knowledge and to compete in business.

Alexander and Foley (1999) have worked out the & Green Future' in an illustrated article in Time. Some specific ideas are telecommuting with our computers, cars will drive on clean fuel cells, fruits and vegetables will grow on nearby organic farms and energy will be delivered from renewable sources.

Food and agriculture

The public concern for sustainable development deemed as the strategy to meet the needs of the present without compromising the ability of future generations to achieve their own needs (Anon., 1987) is still growing.

The global food security and the sustainable resource management are key issues for agricultural production.

About 840 million people in developing countries do not get su$cient protein and energy (Brown, 1999). Resource management is linked to land use, energy and water. The use of energy from fossil fuels is related with the emission of greenhouse gases. At the Kyoto conference, a reduction of the emission of greenhouse gases for Europe for the year 2010 of 6% in comparison with the year 1990 was determined. From several governments, support is given to the use of renewable energy sources as wind, solar energy, biomass and geothermal energy. It may be noticed that during the 20th century the use of energy from renewable sources was decreased by approx. 23% (Flavin & Dunn, 1999). The present consumer is influencing not only the product (shape, quality and safety), but also the production method. In the food production chain, there is more and more the necessity for the control of quality, also in terms of absence of toxic compounds. However, our food was never so good and so safe. Nevertheless, the consumer requires against the background of the dioxin contamination more guarantees for food-safety.

TRENDS IN SCIENCE AND TECHNOLOGY

General

Especially in agriculture, we are confronted with fundamental changes, e.g. the reversal of the chain (consumer/user is governing instead of the producer) and the already mentioned turn-around to sustainable systems (environment, energy, water, agrochemicals). This implies that the generation of knowledge, technology development and innovation play a crucial role in the processes of transformation. In the frame of environmental loads related to agricultural activities, e.g. energy use, water use, emission of greenhouse gases, waste disposal, use of agrochemicals, &breakthroughs' in knowledge, technologies and innovation are necessary. In a &high-tech' information society, we have observed breakthroughs in human mobility, biotechnology and computerization, as realized through mechanization for farm labour productivity. Computerization increased productivity in industry and services.

Knowledge generation, technology development and innovation must be distinguished. These three elements are important for the agricultural engineering discipline.

Scientists must not only have analytical skills, but also the competence for the design or redesign of, for example, biosystems contributing to the development of new food chains. Knowledge generation implies new facts, theories and insights, such as the formation of volatile gases in animal slurry, whereas dielectric measurement principles for the water content in soil or concrete can be denied as a technology. Innovation is a concept, which means more than expertise and knowledge. Innovation is related to new products, processes or concepts for integrated systems and is market oriented (Verkaik, 1997). We also have to distinguish between the explicit knowledge origination in the world of universities and the tacit knowledge. The latter type of knowledge is subjective and connected to experience, context and practice. Therefore, participation in challenging, innovative projects must be broader than between scientists and manufacturers of agricultural machinery only. The success of innovations depends also on the participation and commitment of, for example, growers, auctioneers, supermarket managers and bankers. Innovation of products, processes and systems is as important for manufacturers of agricultural machinery as to earn and secure the licence to produce for the farmers and growers (Brega, 1998). In the Netherlands, a special programme is developed for sustainable technology development (STD). The goal is to achieve an efficiency, that is 20 times better than today. In other words, the ambitious goal is the reduction of the environmental strain by a factor of 20. The central issue of this programme is: &No prosperity without sustainability.'

The focus is on completely new technologies or combinations of them, that push back the frontiers of science and technologies (Anon., 1997). Networking between different stakeholders in industry, institutions and agriculture is necessary to achieve the ambitious goals in projects on food production, sustainable land use and use of solar energy.

Besides scientific developments in technology fields, it is interesting to be aware of the central challenges of the neurobiology of cognition. For the goal of adding intelligence to agricultural machinery and equipment, it is important to follow discoveries and growing insights in understanding of the higher brain functions (Nichols & Newsome, 1999). Also for the architecture of decision support systems, the brain function is an exciting boundary. Professionals in agricultural engineering fields have not always foreseen possible elects of innovative concepts or systems. In the larger part of European countries, criteria such as improvement of production efficiency and scaling-up of farm enterprises have no longer priority in the society. Illustrative is the example of automatic milking systems, on which a technology assessment procedure is carried out. One of the results of the procedure was the special point of combining the automatic milking system and grazing. Also farmers must be aware of working with an automatic milking system. Complex innovative technologies require a technology assessment procedure, e.g. automatic picking of fruit vegetables. This is also infuenced by the changing relation between science and society. Gibbons (1999) is convinced that in the 21st century knowledge must be socially robust. Socially robust knowledge is that which is evaluated, validated, tested and retested in many more conditions and situations. The current debate on genetically modified organisms is an example that the discussions on food safety and risks must be carried out with larger groups of the community. Science today needs to enter a dynamic process where it must be legitimized again and again.

Technology foresights

Which technologies are strategic for the agricultural business in general and the agricultural engineering field in particular? A technology would be considered strategic if it has a strong potential for creating opportunities than can be protect ably exploited across the spectrum of business and industry in the Netherlands. In the technology chain basic research, applied research, technology development and technology use play a role. In the mentioned report, biotechnology (breeding of plant and animals, gene technology), discrete production (production automation technologies) and information and communication technologies (data and knowledge systems, computer-aided design, computer modelling and simulation) are stated as strategic linked to the agribusiness and fishing. These technologies are related to the system development in precision farming, quality improvement of products and reduction of environmental load. From nine technology clusters, including materials technology, biotechnology, and energy technologies, with 46 technology "elds (e.g. surface treatments, bioprocess technology, renewable energy technologies), 15 technology fields were classified as strategic. From those 15 technologies, the bioprocess technology, data and knowledge systems, energy saving technologies, gene technology, measurement and process control, mechatronics, microelectronic components (sensors and actuators), production automation technologies, software engineering and separation technologies are strategic for the agricultural engineering discipline. In comparison with other technology foresight studies in the USA, UK, Germany, France and Japan with a time scale from 5 to 20 years, some additional strategic technologies were identified, such as nanotechnology (electron/ion microbeams with relevance for miniaturization) and computer and network systems. Besides the economic and specific criteria, social criteria also play a role in identifying strategic technologies in the last- mentioned technology foresight studies.

Challenges for the agricultural engineering discipline

Kitani (1994) describes the dilemmas of population, poverty and pollution and economical development, energy and environment. There is an increasing need for food and feed in the world, feed because of the fact of the growing demand for meat in the Asian countries. The issue of energy in agriculture has a strong relation with the knowledge and application of knowledge of the agricultural engineering discipline. The higher quality of life desired by mankind is observed by the growing demands for food, feed, fuel and energy. Kitani (1994) supports the widening of the horizon into a broader area of biop roduction instead of expanding the traditional livestock production. The role of agricultural engineering is to support bioproduction through meteorology and environmental technology support and expertise in the field of energy, power, work-study and ergonomics. Important trends in Western Europe's agriculture are the shift from producer to consumer orientation, changing consumer preferences, globalization (a!ects both agricultural enterprises and markets), mass individualization (custom- made products), competition through knowledge and the formation of new agrochains. An example are the growers of ornamentals which renew continuously their assortment and marketing methods (Van Oosten, 1998). On the basis of the general trends in the society the challenges for agricultural and environmental engineering are the contributions to sustainable agricultural practices. The characteristics of sustainable agriculture must be rethought and &translated' in terms of the different working areas, e.g. ecotechnology, biodiversity, working conditions and quality of products and systems. It must be underlined that in the change to and the development of a sustainable agriculture the combination of economy, ecology and technology is crucial for the social acceptance and support.

In the frame of the European Commission, Sevila and Steinmetz (1994) have published a comprehensive report on the role of agricultural and environmental engineer ing. The mentioned challenges are the flexibility of production systems through diversification, the improvement of quality management on environment, animal welfare, safety, reduction of production costs (competitive), contributions to the improvement of the ecosystem (landscape, watershed, waste recycling, rural employment and the development of technical support (planning and decision support).

The next logical step is the connection of challenges with strategic themes (Jongebreur, 1999a). The trends in the more classic fields of agricultural engineering, namely mechanization and automation, structures and environment, labour and management, energy and water and measurement technology are analysed by Jongebreur and Speelman (1997).

10

ECONOMICS OF AGRICULTURAL MACHINERY

Agricultural machinery is machinery used in the operation of an agricultural area or farm.

STEAM POWER

Power for agricultural machinery was originally supplied by horses or other domesticated animals. With the invention of steam powercame the portable engine, and later the traction engine, a multipurpose, mobile energy source that was the ground-crawling cousin to the steam locomotive. Agricultural steam engines took over the heavy pulling work of horses, and were also equipped with a pulley that could power stationary machines via the use of a long belt. The steam-powered machines were low-powered by today's standards but, because of their size and their low gear ratios, they could provide a large drawbar pull. Their slow speed led farmers to comment that tractors had two speeds: "slow, and darn slow."

Internal combustion engines

The internal combustion engine; first the petrol engine, and later diesel engines; became the main source of power for the next generation of tractors. These engines also contributed to the development of the self-propelled, combined harvester and thresher, or combine harvester (also shortened to 'combine'). Instead of cutting the grain stalks and transporting them to a stationary threshing machine, these combines cut, threshed, and separated the grain while moving continuously through the field.

Combines might have taken the harvesting job away from tractors, but tractors still do the majority of work on a modern farm. They are used to pull implements-machines that till the ground, plant seed, and perform other tasks.

Tillage implements prepare the soil for planting by loosening the soil and killing weeds or competing plants. The best-known is the plow, the ancient implement that was upgraded in 1838 by John Deere. Plows are now used less frequently in the U.S. than formerly, with offset disks used instead to turn over the soil, and chisels used to gain the depth needed to retain moisture.

The most common type of seeder is called a planter, and spaces seeds out equally in long rows, which are usually two to three feet apart. Some crops are planted by drills, which put out much more seed in rows less than a foot apart, blanketing the field with crops. Transplanters automate the task of transplanting seedlings to the field. With the widespread use of plastic mulch, plastic mulch layers, transplanters, and seeders lay down long rows of plastic, and plant through them automatically.

After planting, other implements can be used to cultivate weeds from between rows, or to spread fertilizer and pesticides. Hay balers can be used to tightly package grass or alfalfa into a storable form for the winter months.

Modern irrigation relies on machinery. Engines, pumps and other specialized gear provide waterquickly and in high volumes to large areas of land. Similar types of equipment can be used to deliver fertilizers and pesticides.

Besides the tractor, other vehicles have been adapted for use in farming, including trucks, airplanes, and helicopters, such as for trans

Technology

Only with mass production being aided by modern technology and intensive marketing can the agriculturist exploit both the domestic market as well as the international market to the fullest extent. The volume of production depends not only on the capital investments and marketing strategies but also on the technical capacity used during the production and processing stage.

In fact, technology has come to play a very significant role even in marketing these days. Technology is absolutely critical to the agro industry be it at the primary (production), secondary (processing) or tertiary (marketing and packaging) stage. Experts have always suggested that agricultural technology could play a vital role in addressing the issues and concerns relating to the conservation and management of rural resources.

Ploughing is the first preparation for planting. The plough is primarily designed to prepare the ground for cultivation by turning it over, thus burying the weeds and loosening the earth. It is generally agreed by historians that the earliest implement used for cultivation was probably a crude pointed bent stick or tree branch which was used to stir the soil surface. In effect, a hand held hoe was used in which the user scratched at the earth to form a tilth where corn could be sown. Over a period of time, these hand held hoes soon developed into simple ploughs. These primitive ploughs were eventually pulled by animals like oxen, camels and even elephants. Animals enabled the land to be tilled more easily and faster; thus more food was produced. The credit for this innovation goes to the Egyptians. These ploughs had different modifications in different parts of the world.

During the period from the 1820s to the 1840s, several innovations occurred in plough production. The Breaking Plough, or Prairie Breaker, was a heavy wooden plough plated with iron strips to reduce friction. Prairie ploughs were heavy, weighing at least 125 pounds and requiring from three to seven yoke of oxen. Cutting only three inches into the soil, farmers could break eight acres a year. After a span of few years, ploughs containing a polished wrought iron moldboard and steel share were invented.

Harrow

After ploughing, other implements were used. The harrow was necessary to smoothen the soil in areas where the soil remained rough. It consists of a wooden or metal framework bearing metal disks, teeth, or sharp projecting points, called tines, which is dragged over plowed land to crush the clods of earth and level the soil. Harrows are also used to uproot weeds, aerate the soil, and cover seeds.

In the beginning the harrows were as simple as a tree branch but the harrow became more sophisticated after the Industrial Revolution. By the 1790s, two distinct types of harrows were in use: the square and the triangle, or "A" frame. The square harrow was used on old fields that were free of large obstructions, while the triangular frame was used on freshly ploughed fields. These models had wooden frames with wood or iron teeth.

In modern times, harrows are of varied types. Some are simply dragged behind a tractor or draft animal; some are suspended on wheels; many have levers to adjust the depth of the cut.

Impact of the Industrial Revolution on Agriculture

Industrial revolution brought about drastic changes in the farming process. Farmers discovered the crop rotation system that allowed them to forgo leaving up to half of the land unused or fallow between each planting. This phase witnessed the use of animal husbandry. The industrial revolution brought about an end to tedious handiwork and encouraged the invention and manufacture of other labor-saving farm implements and machinery. Few of the inventions include:

Seed Driller

Seed drill was an innovation that allowed seeds to be easily planted deep into the earth instead of on top where the majority were washed away or otherwise lost. The machine was pulled by horses and consisted of rotating drills or runners that planted seeds at a set depth.

Horse Hoe

It is horse-drawn machine which loosened the soil and killed weeds.

Reaper

The first reapers cut the standing grain and, with a revolving reel, sweeping it onto a platform from which it was raked off into piles by a man walking alongside. The reaper could thus harvest more grain than five men using the earlier cradles.

The reaper was eventually replaced by the self-propelled combine, operated by one man, which cuts, gathers, threshes, and sacks the grain mechanically. The reaper was the first step in a transition from hand labor to the mechanized farming of today.

Threshing Machines

Prior to the threshing machines farmers used an implement called 'flail' to simply beat the grain with sticks or ropes to knock the seeds from the stalks. But this was a back-breaking work and was of low productivity. Threshing machines were designed for rapidly removing the husks from grain.

With improvements in design and efficiency, threshing machines became progressively more common and the hand flail was gradually consigned to history. The machines could be driven by wind or water power, or by horses, but the steam powered thresher became the most familiar sight. They were eventually replaced in the middle decades of the twentieth century by the combine harvester which both harvests and threshes the crop in the field in a single operation.

Tractor

Tractor is a vehicle particularly crafted to exert traction at slow speeds, for the purposes of hauling a trailer or machinery used in agriculture. The versatility of tractor is with respect to its attachments that it supports. The more the options for connecting attachments to the tractor, the higher is the cost. The most common tractor attachments include front end loaders, mowers, box blades, spreaders, tillers, plows, trailers and backhoes for plowing, tilling, disking, harrowing, planting, and similar tasks. Most tractor attachments are interchangeable so they can be used with different machines. The first tractors were steam-powered ploughing engines, followed by Gasoline Powered Tractors.

Agricultural Machinery Industry in India

The country witnessed unprecedented growth in agriculture which has helped India to graduate from hunger to self sufficiency in food grains by increasing the food grain production from 51 million tonnes to 208 million tonnes, with surplus for export. The technology back-up by agricultural scientists, in the form of "Green Revolution" combined with industrial growth, positive policy support, liberal public funding for agricultural research and development and dedicated work of farmers contributed to the phenomenal increase in agricultural, animal and fish production. Application of engineering in agriculture was equally appreciated by the farmers and to-day they feel proud to have improved machinery from Bakhars to rotavators, Persian wheel to drip and micro-sprinkler systems, cone-dibblers to pneumatic planters, sickles to combine harvesters, sieve to colour sorters, and, kolhus to solvent extraction plants, and hand mills to roller flour mills, etc. The farmers are not afraid of hot/cold desert and vagaries of weather as they have green houses and low tunnel plastic houses technology to grow crops in any place at any time of the year.

The growth in adoption of agricultural machinery in the country has be impossible due to their local manufacture. The manufacture of agricultural machinery in India is under taken by village artisans, tiny units, small scale industries, organized medium and large scale sector. Organized sectors manufacture sophisticated machinery such as tractors, engines, milling and dairying equipment. Traditional hand tools and bullock drawn implements are largely fabricated by village craftsmen and power operated machinery by small scale industries.

Historical Perspective of Manufacturing of Agricultural Machinery

Blacksmiths and carpenters have been the traditional fabricators of agricultural equipment in India. The early agricultural mechanization in India was greatly influenced by the technological development in England.

In 1889, Watts and Kaisar introduced ploughs, corn grinders and chaff cutters Cawnpore (now Kanpur) Experimental Farm in Uttar Pradesh. Sardar Joginder

Singh (1897-1946), who was the Agriculture Minister in the Punjab Government (1926-37), introduced the steam tractors in India in 1914 for reclamation of waste land and eradication of 'Kans'. Horse drawn and steam tractor operated implements were imported during the latter part of the 19th century. The horse drawn equipment imported from England were not suitable for bullocks and he-buffaloes used in India and thus, were suitably modified by small scale manufacturers to suit Indian draught animals. With the establishment of Allahabad Agricultural Institute, Allahabad in 1942, the development activities in agricultural machinery accelerated and as a result bullock drawn Meston, Shabash and Wah-Wah ploughs were introduced in Uttar Pradesh, manufactured by the Agricultural Development Society, Naini in early forties.

The Indian farmers gradually responded to farm mechanization technology especially after Green revolution in 1960s. High yielding varieties with assured irrigation and higher rate of application of fertilizer gave higher yields and better economic returns. This enabled the farmers to start adopting mechanization. The development of power thresher with integrated Bhusa making attachment and aspirator blower and mechanical sieves for grain and straw separation in 1960s was the major achievement of Indian engineers which was widely adopted by our farmers. Gradually demand for other farm machinery such as reaper and combine harvester also increased. Demand of tractors in the country was met through importation until 1961 when Eicher Tractors Ltd. and Tractors and Farm Equipment Ltd started manufacturing tractors with foreign collaborations. To meet the additional demand , importation continued up to 1977. Meanwhile many other industries started manufacturing tractors with foreign know how such as Gujarat Tractors Ltd (1963), Escorts Ltd (1966), International Tractors (India) Ltd. (1966), and Hindustan Machine Tools Ltd (1977).

Punjab Tractors Ltd. started their production with indigenous technology in 1974. Many more industries started manufacturing tractors since then with indigenous and foreign know how.

FARM POWER IN AGRICULTURE

Unit farm power

During early sixties, a concept of farm power availability per hectare basis was used to indicate level of farm mechanization. The potential power availability was used as the measure and could not reflect critical constraint of farm power availability during peak requirements or the actual level of use. The total farm power availability from animate and mechanical sources in 1951-52 was 0.20 kW/ ha which increased to 1 kW/ ha in 1996-97. Animate power contributed 60% of the total farm power in 1971-72 and mechanical and electrical together contributed only 40%. In 1996-97 the contribution from animate power reduced to 21% and from mechanical and electrical power it increased to 79%. The farm power input per unit cultivated land in India is still very low compared to South Korea of 7 kW/ha, Japan 14 kW/ha and United State of America of 6 kW/ha. It is evident that mechanical contributed about 78% to the total farm power but for tractive power it is only less than 30% and thus major use of mechanical power has been for stationary farm operations only. The growth in the ratio of availability of mechanical power to total power availability (per ha basis) and mechanical tractive power to total power availability in Indian agriculture indicates that although 78 per cent of farm power was available from mechanical power sources in 1996, only 29.6 per cent of mechanical tractive power offering higher level of technology was available to Indian farms.

Energy ratio for mechanization indicator

A major defect in considering power in quantifying mechanization is that it does not reflect the dimension of time and thus does not bring in the actual scenario in focus. Energy, from this point of view, is a better measure as it can define the actual quantity under use or, in other words, the potentiality of use of available power sources in real terms.

Based on use pattern of diesel, electricity, human and animal power in Indian agriculture, the trend in use of operational direct energy and its projection for the year 2000. It may be seen that the total animate energy availability is decreasing over the years from 2937 MJ/ha in 1971 to 2505 MJ/ha in 1996- 97 due to decrease in availability of animal energy.

Human energy use increased from 1,331 to 1,525 MJ/ ha between 1996-97 and 1971. The animal energy use, however, decreased from 1,606 to 980 MJ/ha, indicating a negative annual growth of 1.9 per cent. Increased cost of maintenance of draught animal has been a main cause of this trend. Energy derived from electrical and diesel fuel used by mechanical power sources continued to increase with growth rate of 10.7 and 11.1 per cent respectively to supplement the additional energy need for increased land productivity. The total energy use for crop production increased from 3,282 to 8,773 MJ/ha (annual growth rate of 3.6 per cent) during the period.

The share of mechanical energy over total energy used in crop production has increased from 11 per cent to 76 per cent during the same period. In order to meet the operational energy needs, the contribution of mechanical energy has to be increased substantially. To encourage use of tractors in agriculture, more diesel is required for the purpose. In order to achieve the production targets of food grains from 125 million hectare land, and assuming that the available energy is used for its production, the average specific energy availability on all India basis with present growth rate would be of the order of 5860 MJ/ tonne of food grain as compared to 5480 MJ/ tonne in 1996-97.

Farm Power Sources in Agriculture

The use of farm machinery depends upon the farm power sources available in the country for various tractive and stationary operations. Human and animal power, the two 'renewable energy' or ' bio-energy' sources, have traditionally been used for various farm operations. The crops are protected from pests, diseases and from weeds, through the application of chemicals.

The application of these inputs is achieved through 'human power' in traditional agriculture. Animate power contributed 60.37% of the total farm power in 1971-72 and mechanical and electrical together contributed only 39.63%. In 2001- 02 the contribution from animate power has reduced to 16.38% and mechanical and electrical power, it increased to 83.62%.

This change in the relative contributions of the animate and nonanimate power sources has been due to the requirements of modern agriculture where timely application of crop production inputs is an important factor to maximize return on input investments. The nonanimate power equipment (like tractors, power tillers, self-propelled planting and harvesting machines, diesel engines and electric motors, etc) also helps farmers to increase cropping intensities through faster turnaround time between successive crops. In addition, the mechanical/Electric machines reduce the drudgery involved in the traditional methods of farming.

Agricultural Tractors

The sale of Agricultural tractors and other farm equipment has increased. Today more than 250,000 tractors are manufactured every year by 13 manufacturers. These tractors are available in different horsepower ranges of less than 25 to more than 55 horse power. The share of various HP tractors in the recent years has been as follows.

Different sizes of tractors are manufactured in India ranging from less than 25 HP to more than 45 HP but most popular range is 31-35 HP.

The Tractor sales show that their demand is region specific. Punjab, Haryana and western UP constituted the major Tractor market in 1980's with 55-57% share of total All India sales. With increased Tractor population in these areas and good successive rains in 1990's, coupled with changes in cropping patterns like adoption of more profitable cash crops (Oilseeds, Pulses, etc.), and better prices, the sales in Gujarat, MP and Rajasthan have been seeing good growth. The contribution of these states improved from 20-22% in early 1990's to 30-32% by the close of the decade.

The share of eastern states, namely Bihar, Orissa, West Bengal and Assam had been consistently low at 3-5% due to various socio- economic, agro-climatic and other reasons. The credit availability to the farmers in this area has been another major reason for the slow growth in the eastern states. The tractor sales, since mid 1990's, have increased to about 10-12% of All India Sales.

Farm Machinery Industries In India

The adoption of mechanization technology depends upon the local manufacture and after-sales-services besides credit and financial incentive provided by the Government. The manufacture of agricultural machinery in India is quite complex comprising from village artisans, tiny units, small scale industries to State AgroIndustrial Development Corporations and organized tractor, engine and processing equipment industries.

Traditional hand tools and bullock drawn implements are largely fabricated by village craftsmen and small scale industries. Organized sectors manufacture sophisticated machinery such as tractors, engines, mills and dairying equipment. The small-scale industries seldom have R&D facilities and they depend upon public institutions for technological support. They require not only drawings but also prototypes and technical guidance to manufacture the equipment. These industries however, upgrade the technology with experience.

Classification of Industries

The classification of industries in India is based on total capital investment (plant and machinery) rather number of workers employed. These are (i) village craftsmen, (ii) cottage industries, (iii) tiny industries, (iv) small scale industries, (v) medium scale industries and (vi) large scale industries. This classification was done to help the small-scale units through incentives and marketing support.

Village craftsmen

Village artisans are the main source of supply and repair and maintenance of hand tools and traditional implements are made by village craftsmen. These include implements and tools like khurpi, spade, sickle, local ploughs, bakhar, sowing devices, yokes, patela, leveller, oil ghanis, grinding wheels, hand mills, handoperated milk churning tools, winnowing devices, sieves, wooden storage structures, bullock carts, manual water lifting devices etc. If village artisans are properly trained they will accelerate the adoption of mechanization inputs due to their proximity with farmers.

Tiny and small-scale industries

The tiny and small scale units fabricate bulk of improved agricultural machinery such as ploughs, cultivators, disc ploughs and harrows, seed drills, planters, plant protection equipment, reaper harvesters, combine harvesters, threshers, cleaners, graders, mills, crushers, oil expellers, diesel engines, irrigation pumps, dairy machinery etc.

Agricultural machines are reserved for small-scale units. There are more than 18000 such units scattered all over the country but have concentration in selected regions.

Some of these units also fabricate implements and equipment for tractor and power tiller manufacturers. They may lack good machine tools and heat treatment facilities. Some of them are more organized and have better fabrication toolings nd thus are able to manufacture better quality machinery. The bulk of the farm machinery is made by the small-scale industries. They use materials from mild steel to medium carbon steel. Heat treatment practices are generally inadequate except in few industrie manufacturing knife & tillage tools . Equipment manufactured by the SSI units includes Soil working tools, seeding & planting equipment, hand hoes, sprayers & dusters, harvesting & threshing equipment, like reapers, threshers, combines ,maize shellers , decorticators, cleaners, graders, mills, oil expellers etc.

Organized farm machinery industries

The medium scale and large scale industries manufacture diesel engines, electric motors, irrigation pumps ,sprayers and dusters , land development machinery, tractors, power tillers, post harvest and processing machinery and dairy equipment. There are 13 tractor, 2 power tiller, 200 diesel engine, 600 irrigation pump, 48 combine and 188 earthmoving machinery manufacturers . The marketing of agricultural machinery by these industries is through their network of dealerships and, therefore, these manufacturers are able to provide effective after-salesservice. These industries upgrade their product and process technologies through their own R&D efforts, in addition to technological support from external agencies.

Today, India is recognized as a leading country in the world for the development and manufacture of agricultural implements and equipment. The range of equipment includes, tractors, harvesting and threshing equipment, plant protection machines, irrigation and drainage pumps, sprinkler systems, land development machinery, dairy and agro-processing equipment, etc.

India is the exporting increasing volumes of these to various countries including USA, Africa, Asia, etc.

Standardization and quality

The Bureau of Indian Standards (BIS) with its network of centers and laboratories in the country is mandated to ensure quality manufacture and marketing of agricultural and industrial products in the country. The BIS formulates specifications of agricultural machinery and other appliances and prescribes test codes. The Bureau also issues ISI quality certificate marks to the products which meet technical specification as per BIS standards.

The Government of India has established Farm Machinery Training & Testing Centres for promotion of quality farm machinery. In case of agricultural machinery, requirement of quality certification is limited to the sale of agricultural machinery financed under Government Schemes. On certain items connected with safety and health hazards, it is mandatory to have minimum safety standards built into the design or in the installation of machinery during operation. Threshers, chaff cutters, sugarcane crushers etc. fall under this category. The agricultural machines manufactured by the organized sector like tractors ,earth moving machinery, irrigation equipment, plant protection, dairy equipment, processing machinery etc. are certified for their quality by BIS.

Testing and evaluation

The adoption of agricultural machinery is greatly influenced by the quality and after sales available to the farmers. Since manufacture of agricultural machinery is reserved for small-scale industries, the quality is affected by the manufacturing technology adopted by them. Testing and evaluation helps in up-gradation and quality production of machinery. R&D institutions and quality certification agencies conduct the T&E. Testing and evaluation is conducted on newly developed equipment and during its serial production, to facilitate and ensure quality, reliability, durability, functional ease, comfort in operation and cost of operation. Testing is conducted with well defined standard parameters, defined in BIS, ISO, or OECD standards, and where as evaluation is done to measure the performance under simulated or field conditions for the parameters for which the equipment has been designed.

Future Thrus T S in Agricultural Mechanization

India is a large country with wide-agro ecological diversity having predominance of rainfed agriculture, with irrigated agriculture limited to 34% only. Farm holdings are small due to higher population density and land fragmentation will continue due to 'Laws of Inheritance' and 'Hindu Succession Act'. Majority of the farmers have limited surplus money to modernize farms or to invest in improved inputs. Draught animals and increasing agricultural workers population may remain to be the major source of farm power for soil manipulation and for crop handling, particularly in Hill and Mountain regions. Mechanical power for tillage, irrigation, harvesting and threshing will be preferred, including on custom hiring basis. As a result of GATT agreement, prospects of agro-export are likely to increase and product quality standards stipulated under WTO would encourage more and more farmers to adopt modern agricultural production technologies. The future agricultural mechanization technology package therefore may have to;

- be eco-friendly utilizing land water and bioresource catering to the varied group of farm holders,
- facilitate farming operations which are arduous and hazardous,
- increase productivity and conserve resources through effective utilization of chemical, biological and mechanical inputs, and
- modernize commercial agriculture to facilitate agro-export.

Keeping above objectives, the mechanization policy may have to be distinctly different to serve hill agriculture, low lying water logged soils, rainfed and irrigated lands and regions having agro export potential.

Equipment for tillage, sowing, irrigation, plant protection and threshing have been widely accepted by the farmers in India. Draught animal and human power in India will continue to be used, but these are inadequate to ensure timeliness of agricultural operations. Even farmers with small holdings utilize selected improved farm equipment, including through custom hiring. The future mechanization strategy may have to be based on agro-ecological diversity and economic disparity of the farmers. The present trend in agricultural mechanization is for high capacity machines to be used on custom hiring and for contractual field operations.

Rice mechanization, sugarcane mechanization, cotton mechanization, potato mechanization, horticulture mechanization, green house and covered cultivation, drip and micro irrigation are new emerging areas which need attention of Agricultural Engineering Institutions and industries for their development, production and marketing.

Water is a scarce commodity and in future with increasing demand for more irrigation water, concerted efforts will be needed for controlled application of water through drip, sprinkler and micro-sprinkler systems to economize use of water and improving water use efficiency With the shift in agriculture towards diversification and agri-business, substantial areas will go under horticultural crops. This will also help to export good quality high value agri-products for better returns to farmers and to earn more foreign exchange. The green house technology offers ample scope for increasing productivity particularly of high value cash crops like exotic fruits, flowers and bio-tech plants.

Presently little effort has been made to mechanize hill agriculture, where there is tremendous potential of growing horticultural crops, flowers etc. In future this calls for developing appropriate technologies for mechanization.

In order to enforce quality, reliability and safety in the manufacture of agricultural implements, manufacturing of critical components need to be standardized and encouraged for mass production by medium and large scale manufacturers. Keeping long standing demand of farmers and the Ministry of Agriculture and on the recommendation of the Advisory Committee of the Ministry of Industries, the Union Budget of India 1998-99, announced the exclusion of farm implements and tools from the list of items reserved for manufacture by small scale industries sector to enable the farmers to get benefit of wider range of implements and tools at competitive prices, and with requisite aftersale-service. The decision of the Government of India to de-reserve the manufacture of farm machinery will help the organized sector to bring latest farm machinery technology for accelerated adoption by the farmers. The small-scale industries in turn will adopt the technology for local manufacturing at a much lower cost. This will help the small-scale sector to become more competitive and to enlarge their market size.

A Return of the Threshing Ring?

A Case Study of Machinery and Labor-Sharing in Midwestern Farms

Machinery-sharing provides an alternative for smaller producers to obtain the efficiencies of large farming operations and remain competitive in an increasingly concentrated agricultural industry. This research uses a multiple case study design to examine the motivations for sharing equipment and labor among farms and to better understand how group members handle the transaction costs of sharing. Our case evidence finds that in addition to cost savings, access to reliable labor is an important motivation for participating in a sharing arrangement. Trust and frequent communication among group members helps to minimize the transaction costs incurred from sharing.

A number of market forces favoring larger farm operations are driving U.S. agricultural production toward a larger scale. Among them is escalating input costs, especially machinery costs. Estimated per-farm costs for tractors and other self-propelled machinery in Midwestern states rose 54% in real terms over the past decade, from roughly $4400 per farm in 1998 to $6800 per farm by 2008 (NASS, 2009). Because machinery is a "lumpy input" that must be adopted in discrete amounts, expanding production to take advantage of size economies is one way to overcome the associated rise in fixed costs (Johnson and Ruttan, 1994).

Although many producers have increased their farm size, this strategy is not always appropriate or feasible. Insufficient access to land and capital can limit expansion. Expanding production may increase risk as a result of larger investments in a small set of related commodities (Roe, 2005). Also, there are limits to farm size growth. The weather-dependent nature of farming and the high level of uncertainty involved in farming requires quick decision-making by farm managers, a role that cannot typically be delegated to workers (Johnson and Ruttan, 1994). The prevalence of family farms has been explained as a result of their ability to minimize transaction costs, particularly the costs of monitoring workers.

An alternative strategy to deal with escalating farm machinery and equipment costs is to share machinery. Many farmers occasionally share a piece of equipment or trade a few days labor and it is common for neighbors to help when a farmer is sick or injured. However, there is evidence that a number of U.S. farmers are revisiting the practice of sharing equipment and labor with other producers on a more routine basis. For example, the original idea behind the agricultural leasing firm MachineryLink was to share machinery between farmers over a geographic distance to take advantage of differences in growing seasons across regions.

Machinery-sharing rings and other farm-level cooperative arrangements are more common in Europe and Canada. Studies of these machinery cooperatives have documented machinery costs savings for members as well as several related benefits. These include access to specialized or more efficient machinery; scale economies; an ability to draw on the experience, labor, and ideas of other members; the possibility of labor or task specialization; access to volume discounts on other inputs; risk-sharing; and environmental benefits from reduced input use. Potential drawbacks identified include a loss of timeliness in field operations, decreased autonomy in decision making, more complex management, potential problems with lenders and split lines of credit, and difficulties in unwinding the arrangement.

The research on resource-sharing arrangements in the U.S. is sparse. A few University Extension guides address machinery-sharing as one of several options for controlling equipment costs on the farm (for example, see Edwards, 2001; Weness, 2001). Lawless, Cropp, and Harris (1996) address potential advantages and disadvantages of various legal business structures for multifamily dairy operations in Wisconsin but do not analyze other aspects of these arrangements. Stofferhn's (2004) survey of North Dakota farmers and ranchers found roughly 11% of respondents indicating a willingness to share equipment or labor with a neighbor but did not document examples of the actual practice. Evidence of cooperative agreements between producers for sharing equipment is therefore largely anecdotal and not well understood.

This research seeks to 1) describe the motivations of farmers who have entered into agreements to share machinery and/or labor with one another; and 2) document the ways in which these groups have experienced and addressed the transaction costs (timeliness, group decision-making, moral hazard problems) of sharing. Given the limited amount of information on machinery-sharing groups in the U.S., this research is exploratory in nature. We use a case study approach that complements existing theoretical and empirical research by providing more detailed information about how existing machinery-sharing arrangements function.

The article proceeds as follows. We outline a conceptual framework of the benefits and costs of sharing equipment and labor between farms in the next section. We discuss the data collection methods next. Presentation and discussion of the internal and external economies of scale and transaction costs uncovered in the case studies follow. The final section concludes with a summary of research findings.

A Conceptual Framework for Machinery-Sharing

Participation in a machinery-sharing arrangement involves a tradeoff between access to internal and external economies of scale and increased transaction costs

(Allen and Lueck, 1998; Valentinov, 2007). Internal economies of scale arise from improvements in technological efficiency; that is an ability to produce more output with the same inputs or to produce the same output with fewer inputs (Varian, 1992). Working in a group to share machinery tends to increase the number of acres serviced by the machinery, reducing inputs, and average costs for a given amount of output. Sharing can therefore make newer, larger, more technologically advanced equipment economical. In addition, group members can improve labor productivity by coordinating tasks to reduce duplication and allow for task specialization.

External economies of scale are related to advantages larger farms may have in accessing inputs; obtaining and negotiating terms of credit, storage, services; and marketing and distribution opportunities (Johnson and Ruttan, 1994). For example, larger farms can negotiate volume discounts on inputs (McBride, 2003). Larger farms might be able to attract specialty contracts that pay premiums for delivery of greater quantities of product. Like marketing cooperatives that obtain higher retail prices through quality assurance, smaller farmer groups may be able to successfully coordinate production practices such as planting and harvest times to maximize quality specifications (Sexton and Iskow, 1988).

Transaction costs related to farming in a partnership or group include timeliness costs, monitoring costs, and costs related to group decision-making. Timeliness costs may occur when a field operation cannot be performed at the optimal time (de Toro and Hansson, 2004; Larsn, 2007). Monitoring costs may arise to ensure that other members are careful with the shared equipment (an asset moral hazard problem) or are contributing an agreed-on number of hours of work when labor is shared (an effort moral hazard problem) (Allen and Lueck, 1998; Larsn, 2007). In addition, collective decisionmaking may be costly, particularly when the interests of group members are not well-aligned (Hansmann, 1996).

Allen and Lueck (1998) model the optimal farm organization (family farm, partnership, or corporation) as a tradeoff between gains from task specialization and lower capital costs and increased transaction costs, particularly moral hazard costs. They empirically test their theory by examining the prevalence of organizational forms across production types (different crops, different regions, and over time). Their analysis shows that when production is seasonal and there are many stages to production with few tasks, the gains from specialization are limited while monitoring costs are high. In this type of production such as with grain farming, family farms are the dominant organizational form and partnerships are relatively rare.

Larsn (2007) builds on Allen and Lueck's model to incorporate social norms, peer pressure, and dynamics (repeated interaction) as factors that might mitigate the moral hazard costs involved in partnerships. Her survey of 640 Swedish farms found 62% participating in some form of partnership arrangement. The partnership farms in this study were characterized by a high degree of trust and Lars?n concludes that the perceived moral hazard problems were negligible.

It is important to note that both the benefits and the costs of sharing relative to farming alone will vary with the size and scope of the sharing and the parties involved. Sharing only a combine with a neighbor during harvest may lower machinery costs while requiring some amount of coordination between partners but is unlikely to result in volume discounts on other inputs or increased output prices. In contrast, a group of many producers farming in a more integrated manner year-round might encounter both significantly greater benefits as well as significantly greater coordination costs.

Study Design and Data

We examine the tradeoffs between internal and external economies of scale and transaction costs in a set of 10 cases of machinery-sharing groups primarily located in the Midwestern U.S. We focused our study on farms with similar production types, primarily grain production (corn, soybeans, wheat), but chose cases to represent variety in the degree of inter-farm cooperation to understand how changes in the scope of sharing affect transaction costs. Additionally, we sought cases in which some, if not all, members were unrelated, believing that kinship ties might confound the analysis of group dynamics.

Case study approaches are well-suited for gaining an in-depth understanding of current events and for asking the types of "why" and "how" questions at the heart of this study (Kennedy and Luzar, 1999; Sterns, Schweikhardt and Peterson, 1998; Westgren and Zering, 1998; Yin, 2003). Analysis of organizational forms in secondary data like the Agricultural Resource Management Survey is hampered by the fact that most machinery-sharing arrangements are informal and therefore not reflected in the data. Given the relatively small number of machinery sharing groups in the U.S., data collection through a random survey of producers might not identify many, if any, groups for analysis.

The case study approach does have its limitations. We cannot calculate an incidence of machinery-sharing, for example. In addition, our findings do not generalize to the population, but rather to a set of theoretical propositions (Yin, 2003). The case approach taken in this study does allow us to illustrate the range of organizational forms and strategies used to share machinery in Midwestern grain farms, to document differences across cases, and to describe any systematic patterns in these differences. We used a multiple case design in which the unit of analysis was the machinery-sharing group. Potential case study subjects were identified through a web-based survey of University Extension professionals in five Midwestern states. Respondents were asked to identify local groups that fit the following description: "independent farms that are participating in cooperative arrangements to share resources for production, yet retaining decision making sovereignty over their assets and labor." University Extension staff responding to the survey identified 50 groups in five states (Iowa, Illinois, Wisconsin, Nebraska, and Indiana) sharing equipment and/ or labor.

The majority of groups identified in the survey had only two or three members (39 of 50). Half of the groups contained no family members, whereas 30% of the groups were comprised solely of family members and 18% combined family and nonfamily members. Slightly more than half of the groups had existed for more than 5 years. Although the groups were engaged predominantly in grain production, cooperative arrangements among vegetable, livestock, and fruit growers were also identified.

The web survey discovered a wide range of cooperative arrangements. Seventy percent of groups identified share both machinery and labor. Fourteen percent were reported to share only machinery, whereas 10% reportedly share only labor. A variety of equipment types were identified in the respondents' comments. A number of groups share sprayers, combines, and harvest equipment. A few groups share equipment for haying, manure hauling, seeding, and irrigation. Some producers only share the cost of a single piece of equipment. Other groups share whole machinery sets (e.g., combine, tractor, planter, and sprayer) as well as labor for operating the equipment. Some others share not only equipment and labor, but also purchase inputs and market output as a group.

The formality of the sharing arrangements varied from simple verbal agreements to written contracts and formal business structures. Over half of the groups identified have a verbal agreement for sharing resources. Only four of the groups identified were known to have a written agreement. However, for a significant portion of the groups (40%), the type of agreement was unknown by Extension staff at the time of the survey.

Based on the web survey results, we developed the two-dimensional taxonomy of cooperative arrangements shown. The vertical dimension represents the degree of cooperation, ranging from sharing only one piece of machinery to a highly integrated case in which producers share all equipment, purchase inputs jointly, and market at least some output together. The horizontal dimension represents the degree of formality of the agreement ranging from a verbal agreement to a formal business structure.

Cases focused on examples of sharing in Midwestern grain operations, but we attempted to identify cases representing the spectrum of formality in arrangements and degree of cooperation. Selecting cases to represent this diversity was meant to facilitate comparisons among groups. Ten case study analyses based on in-person interviews of producer groups who share or have shared resources among farms were completed between June 2004 and January 2005 It places these cases in the context of our taxonomy. Potential participants were contacted by telephone and by mail to solicit their participation in the study. With the exception of LMC whose members were interviewed during a visit to our university, we conducted interviews with group members "on-site" in a neutral location such as a local Extension office or hotel meeting room in their local community. Once a meeting time and location were agreed on, a pre-interview questionnaire was sent to the group contact requesting basic information about group history and members.

We conducted group interviews with as many group members as possible using interview guides, basic outlines of the general topics, and open-ended questions to be discussed. Interviews lasted approximately 2 hours and each participant received $50 to compensate them for their time. The interviews were taperecorded when possible and later transcribed. Investigators took notes during the interview and wrote summaries of the conversations as soon as possible after the interviews occurred. As a result of a concern that members would not feel comfortable expressing negative opinions about their arrangement in the presence of their partners, we provided each member with an exit questionnaire to be completed at a later time and mailed to the investigators in a stamped, addressed envelope we provided. The exit questionnaire solicited members' opinions about the most negative aspects of their equipment-sharing practices and requested any other information they felt relevant to the study. The notes of the investigators, transcripts, and completed questionnaires served as the basis for the analysis of the case study findings.

The degree of labor-sharing varied from none, in the case of a long-distance equipment-sharing arrangement, to a few weeks during harvest season to fully integrated labor operations year-round. Agreements between group members were of two basic types. Groups involving fewer than four members, regardless of the number of equipment pieces or amount of labor shared, predominantly had only a verbal agreement. Groups with a significant scale and scope of operations typically had a written contract, and most had formed a business entity (e.g., LLC). Nine of the groups were operating at the time of the interviews; one group had dissolved.

Internal Economies of Scale

In nearly all cases, sharing helped groups improve technical efficiency. Sharing machinery and working together reduced per-acre equipment and labor costs in the majority of cases. In others, machine costs per acre remained comparable to farming individually but allowed members access to newer, technologically advanced machinery. Producers reported an increased speed of operations as a result of higher-capacity machines, especially during harvest, fewer breakdowns, larger pools of labor, and more efficient use of labor. As one producer put it, "Three go twice as fast as two." Some producers stated that, as a result of the cost and time savings of their cooperative agreement, they were able to significantly expand the acreage of their crop operation or expand the size of their livestock operation.

Somewhat surprisingly, labor, more so than machinery costs, motivated sharing in a number of cases. Many producers rely on family members or retired neighbors to help during busy times, but finding reliable, skilled, and seasonal labor can be a challenge. Working with nearby producers offers one solution to this problem. Several groups reported that sharing eliminated their need to hire outside labor and improved efficiency of operations, particularly during harvest. Operating as a group allows for some specialization. To the extent that group members can take advantage of their complementarities, sharing can reduce the amount of individual effort required. In several cases, producers reported assuming responsibility for tasks they enjoyed or for which they had special training or knowledge. For example, in the case of Zimmerman and Erickson, Zimmerman takes care of most machinery maintenance and repairs because he is a mechanic by training. Erickson, who has a commercial driver's license, handles much of the hauling and scheduling with their local cooperative.

The exception was the case of Bennett, Taylor, and Nelson. Rather than improving efficiency through group efforts, in this case sharing seemed to reduce it. This failed cooperative effort was stymied by the group's inability to agree on farming practices and scheduling (specifically how, when, and by whom field work would be done) as well as the purchase of an undersized combine for their combined number of acres.

External Economics of Scale

The main external economies of scale reported by groups related to improved negotiating power. Several groups reported advantages with regard to renting land. Panhandle Farms felt their ability to plant and harvest in a timely fashion provided an edge in obtaining crop share leases, which are desirable because they require less capital and carry less financial risk than cash rent leases. The Spaulding family noted an advantage from their group efforts with regard to landlords. They jointly own equipment for tile work, which allows them to make improvements to their rented farms more inexpensively than the going rate.

Another perceived advantage of group size appears to be special treatment from machinery dealers, input suppliers, and local elevators. When the members of Progressive Farmers encountered problems with a new combine, the manufacturer sent two engineers directly to their farms to fix it. They felt this was a direct consequence of their size. The Spauldings reported that the local elevator occasionally extended its hours to accommodate additional deliveries from the group. Anderson and Parker also felt their larger-volume purchases resulted in improved service from their dealers.

In contrast, the cases provided very little evidence of marketing opportunities being exploited. Only Panhandle Farms and LMC jointly marketed any of their production. Only LMC had attempted to coordinate production practices to attract price premiums. LMC expanded its operations into a variety of subsidiaries, including a seed-cleaning business and a specialty crops export venture. Parker and Anderson did suggest that a major benefit of their partnership was having someone to share ideas with: "Two heads are better than one." They felt the opportunity to routinely discuss marketing strategies improved their bottom line, although they continue to market their crop separately.

Even joint input buying was problematic in many cases because members maintained loyalties to different seed and chemical dealers. When groups did report coordinated buying of inputs, they generally claimed savings. For example, the Spauldings reported that coordinated purchases resulted in a discount of $3 per acre on chemicals and several free bags of seed each year. Anderson and Parker estimated a 15-20% savings on seed and chemicals from joint purchases.

Despite the lack of coordination observed in input buying and marketing among these groups, if opportunities arose, these groups would be well positioned to take advantage of them. Their history of successfully coordinating group machinery use gives them experience managing group dynamics other producers frequently lack. As in the case of LMC, and to a lesser extent Panhandle Farms and Zimmerman and Erickson, such improved coordination may evolve over time as the group gains experience working together.

Transaction Costs

Sharing equipment and labor involves transactions costs. Members must coordinate schedules, production practices, and, in some cases, even seed varieties. They must make joint decisions about what type of equipment to share, when to trade, how to operate as a group, how to handle repairs and regular maintenance, and a variety of other issues. The more integrated the operations of a group, the more likely additional recordkeeping is required. These costs seem to be higher in the beginning stages of the sharing arrangement. Once groups have operated for some time, many decisions become routine. Timeliness Costs. Several groups reported that they viewed their separate land holdings as one operation for purpose of deciding which fields to work and when. Groups tried to strike a balance between optimal timing and fairness to individual members. One member of the Spauldings explained their method for making decisions about timing in this way: "Whatever's ready first. I don't think that's ever been a question. Sometimes we go around and do a little bit of everybody's. We don't do all of one person's first." Progressive Farmers uses a more systematic approach. Although their land is not adjacent, it has a general north-south orientation. Initially, the group began planting corn in the north working southward. They would then retrace their path from south to north when planting soybeans. Although this approach seemed efficient, they felt it gave the farmland in the middle an unfair advantage. Now the group uses a circular planting scheme and rotates the starting point each year. During harvest, they combine fields as they are ready while making sure they harvest some of each member's land as they go, not leaving any one member's land for last.

In some cases, differences in land types or locations facilitate sharing. The heart of the longdistance arrangement between the Fergusons and Duncans is their ability to exploit variation in growing seasons between their farms. Their shared combine begins the season on Ferguson's North Dakota farm for small grain harvest. The machine moves to the Duncans' Minnesota farm for corn and soybean harvest on September 20, as specified in their operating agreement. This September 20 date is flexible depending on the harvest conditions in any given year. They try to be flexible and communicate with each other to work out solutions that benefit both parties.

Even neighboring farms can take advantage of differences in their land to mitigate timeliness issues. In Zimmerman and Erickson's partnership, Erickson's property is low and consists of river and creek bottom land, whereas Zimmerman has mostly hill ground. Therefore, Zimmerman's property can usually be worked sooner than Erickson's. They realized that by working together they could avoid some of the timing problems associated with working bottom ground before it was ready and working the upland more slowly than would be desirable. In other words, if the fields were planted as if each producer's land were part of a single, larger operation, the outcome for both producers would be better.

Finally, several groups stated that taking account of one another's decisions about hybrids helped reduced timeliness issues, even if they did not make seed decisions jointly. One producer described it this way: "He doesn't tell me what specific hybrids and I don't tell him what specific hybrids. We know that in general we need to have somewhere in the neighborhood of 1 10 day hybrid on this end of the stick and somewhere in the general neighborhood of 1 12 days on this end of the stick."

Monitoring Costs (Effort and Carefulness). In the majority of the cases, a sense of trust and shared values among partners rendered the effort moral hazard problem moot. Most groups interviewed did not track hours contributed to the group effort. In many cases, they worked together on a daily basis and could therefore "monitor" each other's efforts. Overall, statements such as "You don't ask somebody to do something that you wouldn't do yourself" reflected the sense that all partners were committed to the agreement and worked hard to make it successful.

Two notable exceptions to this finding are provided by the cases of the Spauldings and Bennett, Taylor, and Nelson. At the time of the interview, members of the Spauldings group recognized they would need to devise a system to account for members' unequal contributions of hours. Specifically, one of the younger members ran a growing agriculture-related business in addition to farming. Because the busy time for both operations coincides, contributing his share of hours to the farming operation in the spring and fall was becoming increasingly difficult for him. "He's getting big enough now, you are going to have to make some decisions." Another member, a son-in-law who worked fulltime off-farm and was an inexperienced farm equipment operator, paid $20-25 an acre rent to help compensate for his reduced labor contribution.

In addition to sharing equipment, an important part of the Bennett, Taylor, and Nelson's agreement was shared labor. Going into the arrangement, they thought they shared a similar work ethic and view toward farming. As a result, they did not track hours and did not compensate one another for their time. They encountered problems when Bennett hired his brother to complete fall tillage on his land while he worked with the others harvesting. Because of his fulltime off-farm job, Bennett had limited time off during harvest. Hiring his brother was the only way he felt he could complete both his harvest and fall tillage before he needed to return to his job in town. After harvest, when Nelson and Taylor still needed to complete their fall tillage, the fact that Bennett was not available to help created resentment.

11

NEED OF DRY LAND DEVELOPMENT

Drylands is the common term for three agro-climatic zones: arid, semi-arid and sub-humid, where water resources are limited. Aridity and climate variability are dominant characteristics of drylands. Rainfall is scarce, unreliable and concentrated during a short rainy season, while the remaining period tends to be relatively or absolutely dry. The climates are however sufficient to sustain vegetation and human settlement.

Approximately 40 percent of the world's land area is dryland, encompassing savannah, grassland, woodland and shrub land. Drylands are found in all continents except Antarctica. More commonly recognized drylands include the African Sahel and the Australian outback. Australia, the United States, the Russian Federation, China and Kazakhstan are the countries with the most extensive drylands.

Drylands are a vital part of the earth's human and physical environments. Their ecosystems play a major role in global biophysical processes by reflecting and absorbing solar radiation and maintaining the balance of atmospheric constituents. They provide much of the world's grain and livestock, and form the habitat that supports many vegetable species and micro-organisms. Typical crops in drylands include sorghum, maize, cotton, sweet potato, cassava, coffee, banana, tea and sugar cane.

An estimated 40 percent of people in Africa, South America and Asia live in drylands (UNEP, 2000). The human population of the drylands lives in increasing insecurity as productive land per capita diminishes. Soil degradation in drylands, referred to as desertification, affects or puts at risk the livelihoods of people who are directly dependent on the land for their habitat and source of livelihood. The sustainable development of drylands is essential to achieving food security and the conservation of biomass and biodiversity (UNEP, 2000).

MAIN ISSUES IN DRYLANDS

Desertification, the process of arable land changing into unproductive soil or desert, threatens one-quarter of the earth's land and costs US$42 billion every year (UNEP, 2000). It is caused by climate change, including global warming (UNEP, 2000), and unsustainable land management practices, which result from either inadequate techniques or increasing population pressure, and which lead to land degradation. About 65 percent of all arable land has already lost some of its biological and physical functions (UNSO, 2002) and drylands are particularly susceptible.

There is a link between Poverty and environmental degradation which has been signaled often in the literature of the 1990s (Agarwal, 1989; Dankelman and Davidson, 1988; UNSO, 1994). It was thought that women and men farmers who eked out an existence on marginal lands with little education and no access to agricultural resources, could have been driven to adapt practices that may have harmed the environment. However, futher studies have shown that this relationship is much more complex and there is no direct linear causality between gender, poverty and enviromental degradation. When men and women farmers do not own the land they cultivate, there may also be little incentive for them to make environmentally sound decisions. In additon, lack of access to credit may hamper them from buying technologies and inputs that would be less damaging to natural resources (UNSO, 1994).

As food producers, women and men have a stake in the preservation of the environment and in environmentally sustainable development. Land and water resources form the basis of all farming systems, and their preservation is crucial to sustained and improved food production. A lack of understanding and appreciation of women's and men's knowledge of dryland preservation techniques, as well as a disregard for their priorities as resource users, has led many development interventions to fail or to be rejected by local communities. In view of this, planners are now recognizing the value of learning from women's and men's local knowledge in order to protect and sustain the environment, and are aiming more to ensure their full and equal participation.

The international community's efforts to cope with the earth's deteriorating environment have led to a far-reaching debate on environmentalism and activism, which involves developing and developed countries working at the international, national and local levels. Discussions on gender, environment and sustainable development have been nurtured by increased interest in such problems as deforestation and desertification and by the observation that increasing scarcities of fuelwood, animal fodder and water in certain parts of the world affect women most of all (Braidiotti et al., 1994). Governments, development agencies and civil society organizations are increasingly incorporating a gender perspective in their policies, programmes and projects aimed at protecting drylands. However, the road to achieving sustainable development of drylands is still a long and arduous one, and the commitments made at international conferences and conventions need to be translated into practice, with the full and equal participation of women and men at the local, national and international levels.

The drylands of the world cover approximately 40 percent of the earth's land surface and are a direct source of livelihood for about one billion people, especially in developing countries. However, nearly all drylands are at risk of land degradation as a result of climate change, increasing human population, land over-use and poverty. This represents a threat to the food security and survival of the people living in these areas as well as to the conservation of the biomass and biodiversity.

Drylands pose different challenges for rural men and women because of their different roles, relations and responsibilities, opportunities and constraints, and uneven access and control of resources. Furthermore, agricultural, environmental and related policies and programmes often fail to recognise women's particular needs and crucial contribution in the use and management of dryland resources.

By incorporating a gender perspective in policy, projects and programmes, innovative ways of combating dryland degradation and food insecurity can be promoted, notably through a better understanding of men's and women's roles, and their respective concerns and needs. The result is a more sustainable, relevant and equitable development based on women's and men's full and equal participation, on their respective local knowledge, and on ecological and socio-cultural factors. Such a gender-sensitive dryland development represents a great opportunity for rural men and women to join their strengths to preserve food security and the natural resource base in ways that are sustainable.

Drylands, desertification and poverty

Worldwide, one billion people in 110 nations earn directly their livelihoods in drylands. Nearly all of these people, and the drylands on which they depend, are at constant risk from land degradation/desertification, which can be the result of climate change or natural phenomena but is more likely to arise from human activity.

Desertification and poverty are closely and directly linked to each other. While desertification can lead to famine, malnutrition, under-nourishment, epidemics, economic and social instability and migrations, these can, in turn, cause or increase desertification. In addition, poverty contributes to land degradation in drylands by inducing poor women and men to exploit the natural resource base in an unsustainable manner. Degradation then lowers productivity and incomes, thereby increasing poverty and further exacerbating pressure on the natural resource base.

The survival strategies adopted to combat desertification include the overexploitation of accessible natural resources and migration from rural to urban areas or to other countries, which usually involves men leaving for seasonal or longer-term work elsewhere. [China; India - Telengana, Andhra Pradesh; Niger; Sudan]. The analysis of field experiences shows the link between population growth and land degradation in dryland areas. The advancement of women and gender equality are likely to help reduce fertility rates, thereby having an indirect affect on the sustainable development of drylands.

Gender roles in drylands

There is a lack of reliable sex- and age-disaggregated statistics on roles in dryland management activities. In general, men are responsible for decision-making and the planning of farming activities, while women have little authority and have to seek their husbands' permission before they commit family resources or make decisions. Nonetheless, rural women in dryland areas play a key role in natural resource management and achieving food security. They often grow, process, manage and market food and other natural resources. They are generally responsible for small livestock, vegetable gardens and collecting fuel, fodder and water, as well as carrying out their traditional reproductive roles.

Women's traditional roles (e.g. collecting water, growing food, etc.) are particularly crucial in drylands in terms of natural resource management and food security. Men have usually been responsible for decision-making and planning of farming activities, but they increasingly leave the degraded areas to look for jobs in urban areas, leaving women to assume new roles and responsibilities on the farm. In such a changing context, it is fundamental to be aware of the obstacles hindering full participation of disadvantaged groups, including women.

However, despite their multiple roles in dryland management, women's access to and control overnatural resources (such as land) and agricultural support services (including credit, extension services, etc.) are often restricted. [Burkina Faso, Niger and Senegal; India - Telengana, Andhra Pradesh; Kenya]. This limited access to agricultural resources and services is caused by a series of interrelated social, economic and cultural factors that force rural women into a subordinate role and hamper their productivity, as well as limiting their participation in decision-making processes and development initiatives. In some cases, customary practices and laws that limit women's rights to land prevail over legislation that guarantees those rights. Particularly pressing to the issue of dryland management is the fact that insecure land tenure reduces women's and men's incentives to maintain soil quality because they have no permanent rights to the land. Without secure land rights, farmers have little or no access to credit, rural organizations and other agricultural inputs and services. [Burkina Faso; Burkina Faso, Niger and Senegal; India - Telengana, Andhra Pradesh]

Drylands degradation can lead to changes in gender roles. Such factors as migration, population pressure, education and market forces have resulted in women taking more responsibility. In response to change, for instance when they are left behind in the migration process, women readily assume most of men's traditional roles - in addition to their existing agricultural, domestic and reproductive roles. This extra work and responsibility leads women to demand more equal access to land and fertility control. Consequently, control over resources may change, or women may become increasingly involved in decision-making. Many field experiences illustrate how women and men have taken on new roles to combat desertification (Heyzer, 1995), such as through reforestation and land reclamation activities.

Environmental change frequently has a differentiated impact on men and women and leads to changes in gender roles, with women assuming more work and responsibility.

Given that women and men have different roles in dryland management, the impact of desertification affects them in different ways, and the field experiences suggest that environmental change has a far greater impact on women. For example, deforestation and desertification increase the amount of time that rural dwellers have to spend gathering fuelwood and fodder and fetching water. This is one of the most widely cited examples of the impact of land degradation on women, as many societies traditionally see these as women's tasks. At the same time as women are assuming more tasks and responsibilities, desertification is leading to loss of efficiency in such tasks as cooking (owing to diminishing fuelwood sources) and farming activities (as increased labour is needed to combat desertification). [India - Telengana, Andhra Pradesh; Kenya; Mauritania; South Africa]

The field experiences suggest that an interesting social and economic transition is under way. An increasing number of households in drylands are relying less on agricultural activities for income and more on off-farm employment (as well as on remittances from migrant labour). This is enabling households to meet their food security needs in an environment of declining land productivity and, at the same time, it reduces the pressure on drylands natural resources.

The human dimension in UN environmental and related conventions

Since the early 1980s, considerable attention has been devoted to women's and men's different roles in preserving the environment, and extensive efforts have been made to identify the effects of the international environmental crisis on women. Momentum was reached at the workshop of non-governmental organizations (NGOs) that ran parallel to the first World Conference on Women (Nairobi, 1985), when it was recognized that the themes of "women and development" and "the environment" are interlinked and must be incorporated into policy planning.

Since then, several international conventions and agreements have been adopted by the international community, all including commitments reflecting a broad-based consensus on the need to remove the obstacles to women's and men's equal and active participation in and benefit from development initiatives. The following are the most important of these.

Agenda 21 (UNCED): The UN Conference on Environment and Development (UNCED) in Rio de Janeiro in 1992 clearly acknowledged the important role that women need to play in regard to sustainable development. Chapter 24 of Agenda 21 outlines the necessary increased involvement of women at all levels of decision-making. Around the world and in most cultural and social contexts, women traditionally are responsible for the management of natural resources and the social life of communities.

Convention on Biological Diversity (CBD): The 1992 CBD is designed to protect the planet's biodiversity, including genes, species (plant and animal) and ecosystems. It is based on the recognition that biodiversity is a global asset under threat and commits governments to conservation, the sustainable use of biodiversity and the sharing of benefits. The Convention recognizes women's "vital role in the conservation and sustainable use of biological diversity" and affirms the need for the "full participation of women at all levels of policy-making and implementation for biological diversity conservation".

Beijing Platform for Action: The Platform for Action adopted by the Fourth World Conference on Women in Beijing in 1995 identified the need to involve women actively in environmental decision-making at all levels, and to incorporate a gender perspective in all strategies for sustainable development, as one of 12 critical areas of concern requiring action by states, the international community and civil society. Under Strategic Objective K.1 in the Beijing Platform for Action (1995), governments agreed to encourage the protection, use and promotion of the knowledge, innovations and practices of women in indigenous and local communities, ensuring that they are preserved in an ecologically sustainable manner and that women's intellectual property rights are protected under national and international law (paragraph 253.c; SIDA, 1998).

Convention to Combat Desertification (CCD): The International Convention to Combat Desertification (CCD), adopted in June 1994, expressly highlights the essential role that women play in the sustainable management of drylands. The signatory parties committed themselves to promoting awareness and facilitating the participation of local populations, particularly women, in the decisions that affect them. Although the concept of gender is not specifically mentioned in the convention, it is considered to be an underlying principle (UNSO, 1999). CCD takes equal account of both sexes so as to ensure the participation of women and men in programmes to combat desertification. It emphasizes the need to expand women's opportunities to learn about the conservation and sustainable utilization of natural resources, and addresses the issue of female literacy. Literacy increases women's ability to take part in the conservation and sustainable use of natural resources, giving them greater power to bring their experiences and perspectives to bear on the search for effective answers (GTZ, 1999).

Rome Declaration on World Food Security and World Food Summit (WFS) Plan of Action (1995):Both recognize that full and equal participation of men and women are essential for achieving sustainable food security for all and acknowledge the fundamental contribution to food security by women, particularly in rural areas of developing countries, and the need to ensure equality between men and women. Gender is not specifically mentioned in the objective 3.2. related to combating "environmental threats to food security, in particular, drought and desertification, pests, erosion of biological diversity, and degradation of land (...), and restoring and rehabilitating the natural resource base, including water and watersheds, in depleted and overexploited areas to achieve greater production". However, it is considered to be an underlying principle of the Rome Declaration on World Food Security and the World Food Summit Plan of Action.

The comment is frequently heard that "technologies are available, what's needed is to get them off the shelf and into use." But will the products really sell?

It is more than just a matter of pushing the goods off the shelf and out the door. They need to be designed and deployed with a context in mind, and in consultation with prospective users to meet their needs.

Rather than viewing them as products on a shelf, we find it important and useful to consider how they will enable and motivate the poor to move along 'development pathways' that lead to increasing prosperity, food security, equity and sustainability.

Where are we headed?

As human populations increase, a number of different development pathways could unfold. People might fight over increasingly scarce natural resources and not replace those that are removed, spiraling downward into ever-greater poverty. Or, they could invest in their lands, increasing their production efficiency, incomes and total output, meeting the needs of ever-larger populations.

Both scenarios have been observed in different situations. The research-for-development community has a major role to play in steering development towards the second type of pathway. Such pathways need to have the following characteristics.

Enhance the natural resource base

Unreliable rainfall and nutrient-poor soils are hallmarks of the dry tropics. Without addressing these issues,sustained productivity gains are unlikely. For example, the impacts of improved crop varieties have often fallen short of expectations because these aspects were overlooked or assumed to be somebody else's problem'. A development pathways perspective compels us to keep them in mind.

Consider the circumstances

Possibilities will be greatly influenced by enabling elements in the environment such as policy, governance, infrastructure, services, institutions, education, health care and others. A development pathways perspective considers how prospective steps along the path will be influenced by these elements. For example, if grain prices are kept artificially low by cheap subsidized imports, and fertilizer prices high due to transport and border costs, then high-input cereal cropping may not be a practical step until such constraints can be overcome.

Increase incomes

Even the most effective interventions will not have impact if they are not used. Profitable technologies strongly motivate uptake by poor farmers, as long as the enabling conditions are in place such as affordability, access, and know-how. Profits enable farmers to invest in the next step in the development pathway.

Connect to markets

Markets are the source of the profits that motivate progress along development pathways. Without growing markets, the adoption of more productive practices simply creates produce that cannot be sold at a remunerative price, leaving farmers worse off than before. A development pathway perspective requires that we carefully consider how the increased output of goods and services will be translated into hard cash.

Manage risks

Drought, and marketing uncertainty (prices) are two main risks in the dry tropics. As they increase their investments to respond to market opportunities, farmers may be taking on more risk. A debilitating drought can wipe out years of progress, pushing them back to square one on their development pathway. A range of strategies have been devised to help manage these risks, including diversification, water harvesting, weather forecasting, crop insurance, inventory-credit associations, and others (e.g. see previous 'What ICRISAT Thinks' on drought and on microdosing).

Diversify

Diversification helps spread risk and increase incomes. A wide range of crops can be grown in the sunny, moderate climates of the dry tropics, if drought risk can be managed. Many of these are highvalue crops with strong market potential. Diversification thinking can go beyond crops and agriculture to include off-farm enterprises that are less vulnerable to drought.

Innovate

A degree of despondency often accompanies development thinking about the dry tropics. Assumptions that little can be done need to be challenged. Many of the richest agricultural areas in the world are in dry areas (e.g. California, France). Dryland farmers in the developing world continuously innovate to capture new opportunities. A development pathway perspective should for example consider how these areas might progress towards irrigation development, which can fundamentally transform what is possible.

Customize and adapt

There is no single 'one-size-fits-all' development pathway to be handed out to the inhabitants of the dry tropics. Conditions differ enormously. In some rangeland settings, extensification makes more sense than intensification. If markets cannot be accessed, the pathway needs to focus on food security and self-sufficiency. Within-season, farmers need to adapt to changing weather patterns and markets. Rather than prescriptive recipes or technologies taken off the shelf, the research-for-development community needs to provide prototypes, principles, and knowledge that local communities can choose from, customize and adapt to local trends and circumstances.

Steady and sustained

Development pathways thinking can get us away from the 'quick-fix' mentality that is better suited to short-term project organization, than to long-term sustainable development. Development pathways thinking recognizes that in order to step into the future, one foot has to first be firmly grounded in the present. Farmers are pragmatic; they will first test a new intervention on a small part of their land, gradually expanding only if and when it proves its mettle. This safeguards them against the 'boom-andbust' calamities that have too often resulted from pressures for quick fixes.

A development pathway begins with a single step

Confucius famously said that 'a journey of a thousand miles begins with a single step'. Before taking that step, though travelers need to know where they are headed, and what they need to carry to reach their destination. We in the research-for-development community can better assist them in this planning by adopting a development pathways perspective.

We need to plan how one step forward will enable the next step, and how the obstacles ahead will be cleared. We need to consider what is necessary for a safe journey that reaches the desired destination of sustainable prosperity and security.

Governments, scientists, and NGOs are currently making large efforts to improve productivity in the small-scale farming systems that constitute the main livelihood source in sub-Saharan Africa (SSA), as this is seen as one of the most realistic routes to food security, poverty alleviation, and development in the region (Food and Agriculture Organization (FAO)/New Partnership for Africa's Development (NEPAD) 2002, Diao et al. 2007). Current farming systems are primarily rainfed and non-mechanized, use very low levels of external inputs, and yield as little as 1 ton/ha. A key constraint, especially in the semiarid and dry subhumid parts of SSA, is water availability (Falkenmark and Rockström 2004). Although the conventional cure has been the development of large-scale irrigation, recent attempts to overcome crop water deficits have increasingly focused on small-scale solutions, including rainwater harvesting for supplemental irrigation, conservation farming techniques, and other types of small-scale water system innovations (Falkenmark and Rockström 2004, United Nations Development Programme (UNDP) 2006, Comprehensive Assessment of Water Management in Agriculture 2007). This type of technology appears to have significant potential to stabilize and improve yields in small-scale farming systems (Fox and Rockström 2003, Liu et al. 2005, Kahinda et al. 2007, Makurira et al. 2007), and the interest in water system innovations as an option for semiarid and subhumid SSA is currently high (Peacock et al. 2007).

Dryland agro-ecosystems are, however, inherently dynamic, with rapid rates of social and ecological change, making it difficult to predict their future development (Mortimore 2005, Reynolds et al. 2007, Gordon et al. 2008). The rainfall variability generates large spatial and temporal variations in productivity (Millennium Ecosystem Assessment 2005), and a range of other processes, including urbanization, increasing connectedness, market developments, and changing values, have driven recent changes in traditional production and livelihood systems (cf. Bryceson 1996). The strong pressure for rapid transformation to reach the Millennium Development Goals by 2015 further increases the uncertainty surrounding the future of these agro-ecosystems. If not accounting for these processes of change, there is a risk that current development interventions will be outdated before they are implemented. Past examples of this include the promotion of specific cash crops where the market demand has later declined rapidly. Similarly, "silver bullet" approaches focusing on single technologies have frequently failed, as the rate of technological change has been higher than the rate of adoption. A well-known example is the treadle pump that has been widely promoted by NGOs in both Africa and Asia over the past two decades, but which is becoming redundant because of the small motorized pumps entering the market and filling the need for water lifting with far less labor (Shah et al. 2000). Consequently, to ensure that investments made today to upgrade small-scale rainfed farming in dryland SSA also are useful over a long-term perspective, they should be able to benefit the local communities over a range of potential development trajectories.

What is Scenario Planning?

Scenario planning is a systematic method for creatively analyzing complex futures. A scenario is an internally consistent and realistic narrative about the future of a region that is informed by existing conditions and processes, while also factoring in future potential drivers of development (Gallopin et al. 1997, Raskin 2005). Scenario planning uses contrasts among several plausible futures to highlight relationships between environmental factors, management choices, and system dynamics in a way that can inform decision making (Peterson et al. 2003). Central to the process is identifying key drivers of change for the considered system, important uncertainties, and system feedback processes that could strengthen or weaken particular development trajectories (van der Heijden 1996, Gallopin et al. 1997). Scenario planning can be used to identify future opportunities and threats, and thereby help navigating social-ecological systems along more desirable trajectories (Peterson et al. 2003). An advantage of scenario planning is that it can incorporate a wealth of information from a variety of disciplines, and although dealing with complex system dynamics, the outcome is often easily accessible for scientists as well as policy makers and lay people (Kok et al. 2007).

Scenario-planning exercises can be participatory or expert led to varying degrees (van Notten et al. 2003) and performed from local to global scales, or sometimes across scales (Biggs et al. 2007). Scenarios have been used in a number of expert-led regional-scale agricultural planning exercises, for example in Iowa (Santelmann et al. 2004), Denmark (Tress and Tress 2003), the Willamette River Basin in Oregon (Baker et al. 2004), and the San Pedro River Basin across the USA-Mexico border

(Steinitz et al. 2003). However, few examples exist where participatory scenario planning has been used with the specific aim of improving the robustness of particular development interventions (Wollenberg et al. 2000a). In this paper, we provide such an example by applying participatory scenario planning to the social-ecological analysis of agricultural development in the Makanya catchment. If organized and applied well, this type of scenario-planning process can become a powerful tool for learning, consensus building, and public action, which can make local communities better at preparing for, adapting to, and shaping their future.

Scenario Planning in the Dryland Development Context

There are currently a number of on-going social, economic, and environmental processes, whose nature is dramatic and whose outcome uncertain, driving change in semiarid and dry subhumid SSA. Although there are strong forces at multiple levels working for development, large parts of these regions are marginalized, lacking infrastructure, public services, and market access, and lag far behind the rest of the world in human development indicators (UNDP 2007). The vast majority of the population live from small-scale farming and livestock keeping, and their livelihood security is, therefore, intimately linked with the local agroecological productivity (World Resources Institute (WRI) 2005). The productivity is, however, limited both by the challenging hydro-climate and by land degradation (Millennium Ecosystem Assessment 2005). Over the past decade(s), non-farm income sources have become increasingly important for smallholders in these regions (Ellis 1998, Bryceson 2002).

This development is driven both by push factors, such as climate uncertainty and lack of fertile land, and by pull factors, such as market developments and changing preferences. One reason for the latter is that the investments made in education over the past decades have created a young generation that is much more educated than their parents, and consequently, has different opportunities and desires (Bryceson 1996). Although these regions still are predominantly rural, with less than 30% of the population living in urban areas (UNDP 2007), the urban population is expected to increase to 50% by 2030 (United Nations department of Economic and Social Affairs (UNESA) 2008). Finally, as some of the aftermath from the 1980s' economic crises and subsequent structural adjustment programs has settled, many of the countries in SSA have managed to maintain a high annual economic growth rate at around 5%-6% over the past years (World Bank 2007). Although this is remarkable, it should also be noted that this is the only region in the world where the number of extremely poor people has increased over the past 15 years. The high level of uncertainty about the future, in combination with the poverty-induced vulnerability among the population, makes it particularly important to design current development interventions carefully. In this context, scenario planning can be useful as a tool for exploring how different interventions might play out over different futures.

Driving Forces for Change and Scenario Starting Points

The scenario-planning process identified 43 drivers of change in the Makanya catchment, which were classified into 10 categories. Access to new agricultural technology was perceived as the most important factor driving development in the area, followed by health and local governance. The scenario participants were quite optimistic, appraising most factors as becoming "better / more" in the coming two decades. This tendency toward optimism has been identified in scenario-planning processes elsewhere (cf. Wollenberg et al. 2000b). However, an important exception to the optimism was that workshop participants expected the future climate to worsen. The factors that were identified as both important and uncertain, and thus selected to form the scenario starting points, were "health," "local governance," "work ethics / family values," "national economy," and "climate". To make sure that the starting points were not overly dominated by local driving forces, which participants were more familiar with, but which may not always be as important, we chose "national economy" over "approaches for collective action," although the latter was perceived as more uncertain.

The Scenarios

The four scenarios, which all are based on the present-day situation in Makanya, but influenced by different combinations of changes in local governance, economy, climate, health, and values, are presented below. We chose this focus both because it allowed us to show several different aspects of the livelihoods simultaneously, including infrastructure, trade, agriculture practices, and environment, and because the crossroad serves as a strong symbol of change.

Agricultural advance

In this scenario, a stable climate, a strong national economy, and investments by Tanzanian businessmen in food-processing industries drive a transition from subsistence-based to market-oriented production systems. Farming becomes mechanized and intensified, and new agricultural technologies spread fast from businessmen to farmers and among farmers. Although the catchment's economy and infrastructure grow quickly, social inequality increases, and the farmers become more vulnerable to crop-price fluctuations. The development improves the wellbeing of many farmers, but it also damages downstream water quality and reduces downstream water availability.

Managing on the margins

In this scenario, a drier climate makes farming-based livelihoods difficult.

In combination with institutional failure, this leads to societal fragmentation, as the weak leadership at local, regional, and national levels fails to deal with the situation. As a response to the declining productivity, a few entrepreneurs switch from farming to other income sources, and manage to improve their livelihoods. However, most of the population is "managing on the margins," forced to extractive use of farmland and other natural resources, and there is a vicious cycle of declining farm productivity and declining quality and quantity of the surrounding natural resource base that traps many farmers in poverty.

Community cohesion

In this scenario, people come together to address shared problems. Effective local leadership is complemented by the activities of NGOs, who enable village development projects. People still rely on small-scale farming, but incomes rise due to an increasing focus on high-value organic production and new income opportunities such as ecotourism. Living standards gradually improve in the community, but livelihood security is still closely linked to rainfall. People work together to reduce the impact of drought, for example by building communal grain storage silos. The agroecological productivity in the catchment is sustained through improved farmland and ecosystem management, but overall the development is considerably slower than in "Agricultural advance."

Industry imperial

In this scenario, the national economy is growing, and national as well as foreign investors are attracted to Makanya to establish new enterprises. In most of the catchment, agriculture loses importance as people engage in mining or start working in factories. Farming persists in the highlands, where irrigated vegetable production remains common. Some of the villages in the catchment grow together, forming smaller towns, and people adopt a more urbanized lifestyle. The development is rapid and living standards improve for many, but the society also becomes more unequal. A substantial part of the money generated in the catchment leaves the area. New environmental problems are emerging, especially related to the mining industry, which pollutes air and water.

Although the four scenarios display fundamental differences, there were also a number of recurring themes that dominated the discussions during the scenario workshop, and that appeared in all scenarios regardless of the starting point. Among these is the strong identity as farmers and livestock keepers held by the people in the catchment (the loss of which in the industrialization scenario was seen as problematic by many of the participants), and the preference for collective action over individual initiatives for development. The latter fits well with the important roles played by central government and NGOs in most of the scenarios.

It was, for example, clear that many participants were uncomfortable with both "Agricultural advance" and "Managing on the margins" specifically because individual entrepreneurship played a more important role in these scenarios than collective action did. Despite these feelings, it was the general opinion among the workshop participants that this development was quite plausible. Finally, there also seemed to be a substantial worry about loss of local customs and changing ideals among the younger generations.

12

DRILLING AND BORING FOR IRRIGATION

Drilling is a cutting process that uses a drill bit to cut or enlarge a hole in solid materials. The drill bit is a multipoint, end cutting tool. It cuts by applying pressure and rotation to the work piece, which forms chips at the cutting edge.

Process

Drilled holes are characterized by their sharp edge on the entrance side and the presence of burrs on the exit side (unless they have been removed). Also, the inside of the hole usually has helical feed marks.

Drilling may affect the mechanical properties of the workpiece by creating low residual stresses around the hole opening and a very thin layer of highly stressed and disturbed material on the newly formed surface. This causes the workpiece to become more susceptible to corrosion at the stressed surface.

For fluted drill bits, any chips are removed via the flutes. Chips may be long spirals or small flakes, depending on the material, and process parameters. The type of chips formed can be an indicator of the machinability of the material, with long gummy chips reducing machinability.

When possible drilled holes should be located perpendicular to the workpiece surface. This minimizes the drill bit's tendency to "walk", that is, to be deflected, which causes the hole to be misplaced. The higher the length-to-diameter ratio of the drill bit, the higher the tendency to walk. The tendency to walk is also preempted in various other ways, which include:

- Establishing a centering mark or feature before drilling, such as by:
- Casting, molding, or forging a mark into the workpiece
- Center punching
- Spot drilling (i.e., center drilling)
- Spot facing, which is facing a certain area on a rough casting or forging to establish, essentially, an island of precisely known surface in a sea of imprecisely known surface
- Constraining the position of the drill bit using a drill jig with drill bushings

Surface finish in drilling may range from 32 to 500 microinches. Finish cuts will generate surfaces near 32 microinches, and roughing will be near 500 microinches.

Cutting fluid is commonly used to cool the drill bit, increase tool life, increase speeds and feeds, increase the surface finish, and aid in ejecting chips. Application of these fluids is usually done by flooding the workpiece or by applying a spray mist.

In deciding which drill(s) to use it is important to consider the task at hand and evaluate which drill would best accomplish the task. There are a variety of drill styles that each serve a different purpose. The subland drill is capable of drilling more than one diameter. The spade drill is used to drill larger hole sizes. The indexable drill is useful in managing chips.

Digging Wells

It's not an easy task to dig a hole down through layers of soil, limestone, sandstone and shale to reach the layers of gravel and sand that catch and store water underground.

Even in earliest human history, it became obvious that the water in surface streams, rivers and ponds would not be enough to supply everyone. So people started digging wells. The Bible says that Moses smote the rock with his rod and fountain of water burst from the ground. Archeological evidence shows us that the ancient Persians, Egyptians and Chinese built wells to tap groundwater.

Those ancient wells were dug by hand, a laborious and dangerous task. As the wells got deeper, the walls were in constant danger of caving in. It could take weeks, months or even years to build deeper and deeper wells.

In the 40s, the technology finally caught up with the need.

Before World War II, augers had been used to drill shallow wells, but there was a limit to how far down an auger could be turned and the walls of the hole were not supported as the well was dug. In another technique, a hollow "drive point" was hammered down through the rock until it hit water. But the drive point percussion system could not produce a very wide hole, and so the amount of water produced by the well was very limited.

The irrigation industry borrowed techniques from the oil industry. Since the mid-1800s, oil well drillers had been building large pyramid shaped derricks over a well site. Then a cutting bit with industrial diamonds imbedded in it was attached to a pipe, and the pipe and bit were rotated and forced down through the dirt and rock. The debris, or "cuttings," were then forced back up the hole by water that was pumped down either the outside of the hole or the inside of the drill pipe. The walls of the well were stabilized by the introduction of a dense clay substance that penetrated the loose dirt on the side walls and hardened it.

Farmers who wanted to drill irrigation wells adapted the oil well technology to their needs. They didn't need a huge semi-permanent derrick. Instead they needed a system that could drill an irrigation well fast and move to the next well site. They built a smaller derrick that would tip up from the bed of a truck. Then they added powerful new engines to drive the drill bits down.

Most well drillers in the Midwest, where water tables are relatively shallow, began using a technique called "reverse hydraulic" rotary drilling. The term hydraulic referred to the fact that they used water to pump the cuttings back up to the surface. A pond of water with clay in it was dug at the surface of the well. The water flowed down around the outside of the hollow drill stem and was then sucked back up to the surface by a pump attached to the drill stem. Gordon Schmidt says that reverse hydraulic drilling was "a very wet process."

Gordon's father-in-law, John Thieszen, built the first irrigation rig in York County, Nebraska, in the 1940s. It was built from a used truck, some new parts and stuff that they scavenged from the farm.

In York County, drillers had to go through 80 to 100 feet of soil and rock to reach a sand or gravel layer that usually contained water. They could tell when they reached sand by the motion and sound of the drill rig - sand is easier to drill through than the rock above it. Also they could look at what was in the cuttings as they came up.

When they reached the top of the sand layer, they kept going. In the 1940s and 50s, Gordon says, they figured that a hole another 60 feet lower in the sand layer would draw enough water for the well. Today, they dig irrigation wells as deep as the sand layer goes, sometimes another 110 or 120 feet below the top of the sand. They have to make sure they take advantage of all of the water potential in the well.

13

SUSTAINABLE FARMING

Sustainable agriculture is the practice of farming using principles of ecology, the study of relationships between organisms and their environment. It has been defined as "an integrated system of plant and animal production practices having a site-specific application that will last over the long term:

- Satisfy human food and fiber needs
- Make the most efficient use of non-renewable resources and on-farm resources and integrate, where appropriate, natural biological cycles and controls
- Sustain the economic viability of farm operations
- Enhance the quality of life for farmers and society as a whole."

Sustainable agriculture in the United States was addressed by the 1990 farm bill. More recently, as consumer and retail demand for sustainable products has risen, organizations such as Food Alliance and Protected Harvest have started to provide measurement standards and certification programs for what constitutes a sustainably grown crop.

Farming and Natural Resources

The physical aspects of sustainability are partly understood. Practices that can cause long-term damage to soil include excessive tillage (leading to erosion) and irrigation without adequate drainage (leading to salinization). Long-term experiments have provided some of the best data on how various practices affect soil properties essential to sustainability. There is a federal agency, USDA-Natural Resources Conservation Service that specializes in providing technical and financial assistance for those interested in pursuing natural resource conservation and production agriculture as compatible goals.

The most important factors for an individual site are sun, air, soil and water. Of the four, water and soil quality and quantity are most amenable to human intervention through time and labour.

Although air and sunlight are available everywhere on Earth, crops also depend on soil nutrients and the availability of water. When farmers grow and harvest crops, they remove some of these nutrients from the soil. Without replenishment, land suffers from nutrient depletion and becomes either unusable or suffers from reduced yields. Sustainable agriculture depends on replenishing the soil while minimizing the use of non-renewable resources, such as natural gas (used in converting atmospheric nitrogen into synthetic fertilizer), or mineral ores (e.g., phosphate). Possible sources of nitrogen that would, in principle, be available indefinitely, include:

1. Recycling crop waste and livestock or treated human manure
2. Growing legume crops and forages such as peanuts or alfalfa that form symbioses withnitrogen-fixing bacteria called rhizobia
3. Industrial production of nitrogen by the Haber Process uses hydrogen, which is currently derived from natural gas, (but this hydrogen could instead be made by electrolysis of water using electricity (perhaps from solar cells or windmills)) or
4. Genetically engineering (non-legume) crops to form nitrogen-fixing symbioses or fix nitrogen without microbial symbionts.

The last option was proposed in the 1970s, but is only recently becoming feasible. Sustainable options for replacing other nutrient inputs (phosphorus, potassium, etc.) are more limited.

More realistic, and often overlooked, options include long-term crop rotations, returning to natural cycles that annually flood cultivated lands (returning lost nutrients indefinitely) such as the Flooding of the Nile, the long-term use of biochar, and use of crop and livestock landraces that are adapted to less than ideal conditions such as pests, drought, or lack of nutrients.

Water

In some areas, sufficient rainfall is available for crop growth, but many other areas require irrigation. For irrigation systems to be sustainable they require proper management (to avoid salinization) and must not use more water from their source than is naturally replenished, otherwise the water source becomes, in effect, a non-renewable resource. Improvements in water well drilling technology and submersible pumps combined with the development of drip irrigation and low pressure pivots have made it possible to regularly achieve high crop yields where reliance on rainfall alone previously made this level of success unpredictable. However, this progress has come at a price, in that in many areas where this has occurred, such as the Ogallala Aquifer, the water is being used at a greater rate than its rate of recharge.

Several steps should be taken to develop drought-resistant farming systems even in "normal" years, including both policy and management actions: 1) improving water conservation and storage measures, 2) providing incentives for selection of drought-tolerant crop species, 3) using reduced-volume irrigation systems, 4) managing crops to reduce water loss, or 5) not planting at all.

Indicators for sustainable water resource development are: Internal renewable water resources. This is the average annual flow of rivers and groundwater generated from endogenous precipitation, after ensuring that there is no double counting. It represents the maximum amount of water resource produced within the boundaries of a country. This value, which is expressed as an average on a yearly basis, is invariant in time (except in the case of proved climate change). The indicator can be expressed in three different units: in absolute terms (km3/ yr), in mm/yr (it is a measure of the humidity of the country), and as a function of population (m3/person per yr). Global renewable water resources. This is the sum of internal renewable water resources and incoming flow originating outside the country. Unlike internal resources, this value can vary with time if upstream development reduces water availability at the border. Treaties ensuring a specific flow to be reserved from upstream to downstream countries may be taken into account in the computation of global water resources in both countries.

Dependency ratio. This is the proportion of the global renewable water resources originating outside the country, expressed in percentage. It is an expression of the level to which the water resources of a country depend on neighbouring countries.

Water withdrawal. In view of the limitations described above, only gross water withdrawal can be computed systematically on a country basis as a measure of water use. Absolute or per-person value of yearly water withdrawal gives a measure of the importance of water in the country's economy. When expressed in percentage of water resources, it shows the degree of pressure on water resources. A rough estimate shows that if water withdrawal exceeds a quarter of global renewable water resources of a country, water can be considered a limiting factor to development and, reciprocally, the pressure on water resources can have a direct impact on all

Soil

Soil erosion is fast becoming one of the worlds greatest problems. It is estimated that "more than a thousand million tonnes of southern Africa's soil are eroded every year. Experts predict that crop yields will be halved within thirty to fifty years if erosion continues at present rates." Soil erosion is not unique to Africa but is occurring worldwide. The phenomenon is being called Peak Soil as present large scale factory farming techniques are jeopardizing humanity's ability to grow food in the present and in the future. Without efforts to improve soil management practices, the availability of arable soilwill become increasingly problematic.

Some Soil Management techniques

1. No-till farming
2. Keyline design
3. Growing wind breaks to hold the soil
4. Incorporating organic matter back into fields
5. Stop using chemical fertilizers (which contain salt)
6. Protecting soil from water runoff.

Methods in Sustainable Farming

Methods

What grows where and how it is grown are a matter of choice. Two of the many possible practices of sustainable agriculture are crop rotation and soil amendment, both designed to ensure that crops being cultivated can obtain the necessary nutrients for healthy growth. Soil amendments would include using locally available compost from community recycling centers. These community recycling centers help produce the compost needed by the local organic farms.

Many scientists, farmers, and businesses have debated how to make agriculture sustainable. Using community recycling from yard and kitchen waste utilizes a local area's commonly available resources. These resources in the past were thrown away into large waste disposal sites, are now used to produce low cost organic compost for organic farming. Other practices includes growing a diverse number of perennial crops in a single field, each of which would grow in separate season so as not to compete with each other for natural resources. This system would result in increased resistance to diseases and decreased effects of erosion and loss of nutrients in soil. Nitrogen fixation from legumes, for example, used in conjunction with plants that rely on nitrate from soil for growth, helps to allow the land to be reused annually. Legumes will grow for a season and replenish the soil with ammonium and nitrate, and the next season other plants can be seeded and grown in the field in preparation for harvest.

Monoculture, a method of growing only one crop at a time in a given field, is a very widespread practice, but there are questions about its sustainability, especially if the same crop is grown every year. Today it is realized to get around this problem local cities and farms can work together to produce the needed compost for the farmers around them. This combined with growing a mixture of crops (polyculture) sometimes reduces disease or pest problems but polyculture has rarely, if ever, been compared to the more widespread practice of growing different crops in successive years (crop rotation) with the same overall crop diversity. Cropping systems that include a variety of crops (polyculture and/or rotation) may also replenish nitrogen (if legumes are included) and may also use resources such as sunlight, water, or nutrients more efficiently.

Replacing a natural ecosystem with a few specifically chosen plant varieties reduces the genetic diversity found in wildlife and makes the organisms susceptible to widespread disease. The Great Irish Famine (1845-1849) is a well-known example of the dangers of monoculture. In practice, there is no single approach to sustainable agriculture, as the precise goals and methods must be adapted to each individual case. There may be some techniques of farming that are inherently in conflict with the concept of sustainability, but there is widespread misunderstanding on impacts of some practices. Today the growth of local farmers' markets offer small farms the ability to sell the products that they have grown back to the cities that they got the recycled compost from. By using local recycling this will help move people away from theslash-and-burn techniques that are the characteristic feature of shifting cultivators are often cited as inherently destructive, yet slash-and-burn cultivation has been practiced in the Amazon for at least 6000 years; serious deforestation did not begin until the 1970s, largely as the result of Brazilian government programs and policies. To note that it may not have been slash-and-burn so much as slash-and-char, which with the addition of organic matter produces terra preta, one of the richest soils on Earth and the only one that regenerates itself.

There are also many ways to practice sustainable animal husbandry. Some of the key tools to grazing management include fencing off the grazing area into smaller areas called paddocks, lowering stock density, and moving the stock between paddocks frequently.

Several attempts have been made to produce an artificial meat, using isolated tissues to produce it in vitro; Jason Matheny's work on this topic, which in the New Harvest project, is one of the most commented.

Soil treatment

Soil steaming can be used as an ecological alternative to chemicals for soil sterilization. Different methods are available to induce steam into the soil in order to kill pests and increase soil health. Community and farm composting of kitchen, yard, and farm organic waste can provide most if not all the required needs of local farms. This composting could potentially be a reliable source of energy.

Urban planning

There has been considerable debate about which form of human residential habitat may be a better social form for sustainable agriculture.

Many environmentalists advocate urban developments with high population density as a way of preserving agricultural land and maximizing energy efficiency. However, others have theorized that sustainable ecocities, or ecovillages which combine habitation and farming with close proximity between producers and consumers, may provide greater sustainability

The use of available city space (e.g., rooftop gardens, community gardens, garden sharing, and other forms of urban agriculture) for cooperative food production is another way to achieve greater sustainability .

One of the latest ideas in achieving sustainable agricultural involves shifting the production of food plants from major factory farming operations to large, urban, technical facilities called vertical farms. The advantages of vertical farming include year-round production, isolation from pests and diseases, controllable resource recycling, and on-site production that reduces transportation costs.While a vertical farm has yet to become a reality, the idea is gaining momentum among those who believe that current sustainable farming methods will be insufficient to provide for a growing global population.

Off-farm impacts

A farm that is able to "produce perpetually", yet has negative effects on environmental quality elsewhere is not sustainable agriculture. An example of a case in which a global view may be warranted is over-application of synthetic fertilizer or animal manures, which can improve productivity of a farm but can pollute nearby rivers and coastal waters (eutrophication). The other extreme can also be undesirable, as the problem of low crop yields due to exhaustion of nutrients in the soil has been related to rainforest destruction, as in the case of slash and burn farming for livestock feed.

Sustainability affects overall production, which must increase to meet the increasing food and fiber requirements as the world's human population expands to a projected 9.3 billion people by 2050. Increased production may come from creating new farmland, which may ameliorate carbon dioxide emissions if done through reclamation of desert as in Palestine, or may worsen emissions if done through slash and burn farming, as in Brazil. Additionally, Genetically modified organism crops show promise for radically increasing crop yields.although many people and governments are apprehensive of this new farming method.

Some advocates favour sustainable agriculture as the only system which can be sustained over the long-term. However, organic production methods, especially in transition, yield less than their conventional counterparts and raise the same problems of sustaining populations globally.

Ensuring our food security

Since the dawn of civilizations agriculture is one sector that impacts and in turn is impacted the most by environment. Hence sustainability of the human race and this world depends a lot on the environmental friendliness of our agriculture.

India is facing a food crisis thanks to the systematic destruction of farmlands and food production systems over the last five decades through uncontrolled use of chemical fertilisers, pesticides, mono-cropping and other intensive agricultural practices. Instead of looking at the real problem the government is favouring false solutions like genetically engineered (GE) food crops.

Ecological farming is the answer to the problems being faced by agriculture in our country today. It will also keep agriculture sustainable. This form of agriculture conserves our soil and water resources, protects our climate, enhances agro-diversity, ensures biodiversity, meets the demand for food and safeguards livelihoods. In short, it ensures that the environment thrives, the farm is productive, the farmer makes a net profit and society has enough nutritious food.

India has a long history of agriculture. Over centuries, farmers in this country devised practices to keep our farms sustainable. Practices like mixed cropping, crop rotation, using organic manure and pest management kept our agriculture sustainable. But things changed for the worse with the onslaught of a chemical intensive model of agriculture, imposed through the so called Green Revolution in 1965.

It was therefore not surprising when the International Assessment of Agricultural Science and Technology for Development [IAASTD], an initiative of the United Nations and World Bank, concluded that small-scale farmers and agro-ecological methods are the way forward if the current food crisis is to be solved. This initiative involved a three year review of all the agricultural technologies in the past 50 years by around 400 scientists across the world.

The IAASTD said that to meet the needs of local communities, indigenous and local knowledge need to be declared as important as formal science. This is a significant departure from the destructive chemical-dependent, one-size-fits-all model of industrial agriculture. The report also acknowledges that genetically engineered crops are highly controversial and will not play a substantial role in addressing the key problems of climate change, biodiversity loss, hunger and poverty.

Campaign story

Greenpeace is not opposed to science nor is it opposed to finding more efficient farming methods. But we will not stand for the wilful destruction of the soil, water and biodiversity for the gains of corporates. Nor will we let human beings be treated as guinea pigs to test new crops. Keeping this in mind the sustainable agriculture campaign is currently focused on the following:

Fertiliser Campaign: Degraded soil with diminishing fertility, food full of toxins, huge carbon footprints and empty coffers. This is exactly what chemical fertilizers are doing in this country. It's time to shift from these to ecological means of cultivation proved to be successful in several parts of the country.

GE Campaign: Touted as the perfect solution to the food crisis, GE crops will only worsen it. Among other things, they pose a threat to human health and compromise on food safety. GE crops should not be released into the open at any cost.

Sustainable agriculture and fertilizer use

Sustainable agriculture must produce enough food and fibre to satisfy changing human needs while conserving natural resources, maintaining the quality of environment and ultimately leading to community and gender equity. In the developed countries in Europe and to some extent in USA, chemical fertilizer has been held as a major culprit for environmental pollution, especially the nitrate enrichment of groundwater. However, evidence available also indicatesthat, animal slurry and septic cess pools largely contribute to nitrate enrichment of groundwaterand also to environmental pollution with ammonia and nitrogen oxides (NOx). There are othersources of environmental pollution such as exhaust fumes from motor vehicles, flyash fromthermal power plants and other industrial effluents. Developing countries such as India, reelingunder population pressure with no additional cultivable land are forced to increase their fertilizer consumption, which as of today is much less than the actual crop needs. Nevertheless, we must learn lessons from the ill-effects of overuse of chemical fertilizer by developed countries and useit judiciously with a well-planned integrated plant nutrient supply system.

Sustainability of our agricultural systems is of global concern today and many definitions of sustainable agriculture have become available. The five main components of these definitions are:

- production of enough food and fibre to meet the increasing and changing needs of the people;
- conservation of natural resources;
- maintaining the quality of environment;
- achieving community and gender equity; and

- avoidance of regional imbalances

Why do we need fertilizers

In agriculture we convert solar energy into chemical energy, i.e. food, feed or fibre. It is common knowledge that this is achieved with the help of chlorophyll in plants. Keeping at optimum other conditions such as solar radiation, temperature, etc. the more the chlorophyll the higher the biomass and it is likely that more food is produced. However, for doing this plants need adequate supply of 13 elements (N, P, K, Ca, Mg, S, Fe, Mn, Zn, Cu, B, Mo, Cl) from soil, which are known as essential plant nutrients. Fertilizers are the chemicals that supply these essential plant nutrients, mostly N, P and K, which are removed by crop plants in the largest quantities.

In one study at New Delhi 28 kg N, 4.4 kg P, 41 kg K, 4.9 kg S, 5.9 kg Ca, 4.2 kg Mg, 400 g Fe, 100 g each of Mn and Zn and 30 g Cu were removed from the soil for each tonne of wheat grain produced. Similarly from the International Rice Research Institute, Philippines it is reported that 10-31 kg N, 1-5 kg P and 8-35 kg K were removed from the soil per tonne of rice grain produced. When crop yields of 5-8 t grain ha- 1 are taken as with the high yielding varieties of crops, it is not possible for most soils to supply the needed amounts of plant nutrients and that is why fertilizers are needed. Such heavy removal of plant nutrients from soil leads to depletion of soil fertility, which shows up in crop yield decline and lowered factor productivity. For example, in a long-term study at four research centres rice yields declined by 32% in plots receiving nitrogen and phosphate fertilization; the decline in unfertilized plots was 57%. From the dryland rainfed areas where yields are low and very little fertilizer applied, a negative balance between crop removal and plant nutrient application of 6.37 million tonnes yr- 1 (N + P2O5 + K2O) in India has been reported. Thus contrary to lowering the rates of fertilizer application, Indian farmers may have to think of fertilizing for crop production as well as for soil fertility resilience. Thus fertilizer has been and will continue to be the king input in India's achieving self sufficiency in foodgrain production.

Fertilizers and sustainability of agriculture

For the last three decades all over the developed world fingers have been raised on fertilizer, particularly nitrogen, as the number one enemy of sustainable agriculture. One should not be surprised to see this happen if the rates of application are too high such as in Netherlands. Nitrogen is a mobile nutrient both in plants and soils and considering the fact that its efficiency (nitrogen taken up by the above ground portion of crop expressed as percentage of that applied) varies from 30-40% in rice to 60-80% in other cereals, a sizeable amount could be added to the environment as ammonia by volatilization from soil surface, nitrous oxide or elemental nitrogen by denitrification which is not restricted to tropical rice regions but also applies to temperate regions and finally as nitrates by leaching in underground water. The ammonia going in the atmosphere contributes to acid rains, while N2O is involved in depletion of ozone layer. What one generally overlooks is that in the case of nitrogen fertilizers we recycle atmospheric nitrogen which is the raw material for ammonia and urea manufacture. Atleast 30-50% of it is converted into human edible food and about one-third is immobilized in the soil and only the rest goes back to atmosphere either as ammonia or as N2O or N2 after denitrification of nitrates. Phosphates, which are not so mobile in soil and get fixed in the soil as insoluble compounds could also leach from very light soils and may also move with eroded surface soil to surface waters such as lakes and ponds. Within the European Union, the average load of nitrogen in 1990 ranged from 35 kg N ha- 1 in Spain to 195 kg N ha- 1 in the Netherlands and average loads of phosphate ranged from 17 kg P2O5 ha- 1 in Spain to 57 kg P2O5 ha- 1 in the Belgium. So great has been the concern of European researchers regarding pollution caused by fertilizers (and manures) that it has been referred to as a 'chemical time bomb' and to overcome the explosion of this the need for immediate change in the land management amounting to abandonment of part of the agricultural land has been suggested.

Pollution of waters

More than any other aspect of environmental degradation, pollution of surface and underground waters due to fertilizers has received greater attention. Water can be broadly classified into four categories: (i) non-flowing such as lakes and ponds; (ii) flowing such as rivers and canals; (iii) river estuaries and (iv) underground water.

Both nitrogen and phosphorus eutrophication of lakes and ponds leads to excessive growth of aquatic plants (macrophhytes) and algae (phytoplankton), which deplete the water of oxygen and this may lead to death of fishes and other aquatic animal life. Furthermore, algal and decaying algal and aquatic plant tissue can lead to discolouration and bad odour and this affects recreational and aesthetic water uses.

As regards flowing water, phosphorus is the limiting nutrient. When all phosphorus is used, plant growth ceases irrespective of the amount of nitrogen available. Thus only if adequate phosphorus is available, increasing concentration of nitrates will lead to algal and macrophyte growth.

In estuaries, on the other hand, nitrogen controls the growth of algae and macrophytes. In USA nitrates and phosphates are suspected of causing hypoxia of the Dead Zone of Gulf of Mexico. Nutrient-enriched run-off from agriculture is also incriminated in the pfiesteria problem that killed a large number of fish in the Chakaspeake Bay, USA. The recommended level of nitrogen is 0.1 to 1 mg l- 1, while the recommended level of phosphorus is nearly one-tenth of that of nitrogen.

Underground water is the major source of drinking water in India and even in developed countries like USA, 90% of the rural population depends on it for drinking water. The safe limit or MCL (maximum contamination level) established by the US Environmental Protection Agency is 45 mg NO3-N or 10 mg N l- 1. The European Union has fixed MCL limit at 50 mg NO3-N or 11 mg Nl- 1. Levels above this may lead to methaemoglobinaemia or blue baby syndrome. This is due to conversion of haemoglobin to methaemoglobin due to nitrites formed from nitrates ingested with drinking water. Haemoglobin is involved in transport of oxygen in the body, while methaemoglobin cannot and thus the patient suffers from anexia (lack of oxygen). There are some indications that excess nitrates in the human body may react with amines and form nitrosamines that may lead to gastric cancer, but substantive data are missing.

In India the highest rate of fertilizer N is in Punjab, where an increase in shallow well waters from 0.04-6.15 in 1975 to 0.31-13.3 mg NO3-Nl- 1 in 1988 has been reported. A recent survey in Delhi reported a range of 26-150 mg NO3-Nl- 1 in shallow well water, which is very high constituting a warning.

Other sources responsible for environmental pollution

There are a number of sources other than fertilizers which are responsible for environmental degradation. These include livestock and human excrement in rural areas and leaking septic systems, sewage, combustion of fossil fuels in motor cars and other vehicles in urban areas. Industries that use nitrates in their manufacturing processes such as meat curing, production of explosives including fire crackers, heat transfer fluid, etc. may also release substantial nitrates in their effluent water. Precise estimates on all these are not available. The available information is briefly discussed.

Livestock excrement

Ammonia emissions from dairy farms are one of the major sources of environmental pollution in Europe. Losses of ammonia occur during slurry application, housing, slurry storage and grazing. On an average 30% of nitrogen excreted by farm animals is released to the atmosphere as ammonia and the bulk of it is contributed by urine. Estimates indicate that 50-90% of N contained in urine is urea and 4-41% of N from applied cattle urine may volatilize. The average ammonia emission per unit land area in Western Europe is 12 kg N ha- 1 yr- 1 and the highest is 45 kg ha- 1 yr- 1; as a contrast ammonia emission from growing arable crops is estimated at 1-2 kg ha- 1 yr- 1. What is to be noted is that ammonia emitted is not deposited in the same area but transported to other areas depending upon wind direction and velocity.

In India where cowdung is mostly used as cooking fuel, loss of nitrogen could be partly as ammonia during storage of wet dung but mostly as nitrogen oxides (NOx) when the cowdung cakes are burnt. Nevertheless it does contribute to environmental pollution. We need to measure and generate data on this.

Ammonia emission is not the only way the animal excreta contributes to environment. After its application to soil, most of ammonia produced is retained by the soil exchange complex and is gradually nitrified. Nitrates are then leached to deeper layers and contribute to underground water pollution. From a study in a rural community in California (USA), based on 15N ratio studies it was concluded that much of the nitrate in groundwater came from sewage or manure. As already discussed, under anoxic conditions nitrates are denitrified to nitrous oxides or elemental N (generally referred to as NOx).

In UK limits of organic manures have been fixed at 250 kg N ha- 1 on grassland and 210 kg N ha- 1on arable land in nitrate vulnerable zones (NVZ); the figure for arable land will be reduced to 170 kg N ha- 1 from 19 December 2002.

Septic systems and sewage

Septic systems and cesspools (a pit or a hole in which sewage collects) in heavily populated areas such as Long Island in USA have been major sources of nitrate in groundwater for decades and public supply wells in the upper glacial aquifer of Nassau county were abandoned in 1949 because of nitrate contamination. Katz et al. estimated that 11.6 metric t N km- 2 were contributed by cesspools and septic systems in the Nassau county in 1975.

Biological nitrogen fixation

Biological nitrogen fixation (BNF) is considered as one of the major ways by which the crop needs of fertilizer nitrogen can be partly (30-60 kg N ha- 1)47,48 met; the rates for green manuring could be 80-185 kg N ha- 1. There can be no two opinions about it, and infact legumes as restorer of soil fertility have been known since ages.

Estimates of total annual terrestrial BNF range from 139 to 170 MT with symbiotic N fixation in arable farming accounting for 35-44 MT N, and another 45 MT from permanent pastures and the rest from non-symbiotic N fixation. In comparison, only 91 MT of fertilizer N was globally manufactured in 1996-97; which is only 53-65% of biologically fixed N. The point to consider is that nitrogen fixed by BNF also enters the soil nitrogen pool and while some of it is available to associated crop, most of it is made available to the succeeding crop. Biologically-fixed N is much more readily available to plants than the native soil organic N, nevertheless, a part of it is also subject to loss. For example, a loss of 28-29% of nitrogen fixed by Crotolaria sp and Sesbania sp has been reported when these legumes were green-manured to rice, as compared to 36% from urea nitrogen. Similarly in maize loss of Leucaena nitrogen was 25-30%, while that from ammonium sulphate was 14-35%. Thus part of biologically-fixed nitrogen may also contribute to environmental pollution including nitrate enrichment of groundwater. BNF is reported to contribute as much as 13.5% of total N_2O emission to the atmosphere, which was equal to the amount contributed by fertilizers.

India's foodgrain needs

In 1951, we were only 361 million people and by 2000 AD, another year from now, we will be 1004.5 million, nearly trebling our population in the second half of twentieth century. A well-planned and concerted effort by agricultural scientists, extension workers, farmers and the government price support policies have made 'green revolution' realizable by increasing grain production at a rate faster than the increase in population. This has kept us away from hunger despite some very bad drought years. Hungry people can hardly think of a quality environment. Three major inputs that have made it possible are: seeds of high yielding varieties of crops, fertilizer and increased irrigation facilities. But then it is not all over and although there are good signs of decline in the rate of population increase from 2.14 to 1.70%, we are still likely to add another 420.5 million people in the next 20 years by 2020 AD, that is about 21 million people each year. This will certainly increase our food demand.

The total foodgrain demand by 2020 is estimated at 294 MT (122 MT rice, 103 MT wheat, 41 MT coarse grains and 28 MT pulses). Thus by 2020 we need to produce about 100 MT of additional food grain yr^{-1} from the same or even less area (some more area will go to meet the increasing needs for roads, rails, buildings, etc.). We have no choice but to increase the fertilizer application. During 1980-90 there has been 3-4% decrease in fertilizer nitrogen consumption in Europe and USA, while in Asia it has increased by 74.4%. In India the current fertilizer consumption is much below the mark. Only 19 out of 437 districts in India consumed more than 200 kg N + P_2O_5 + K_2O ha^{-1} in 1996-97, while 176 districts consumed 50 kg ha^{-1} or less.

The way out

Social scientists are quite right in pointing out regional imbalances and lack of equity among communities due to modernization of agriculture, but fertilizers do not come in the way of achieving either of these social goals. There are other factors responsible such as lack of infrastructure, lack of small scale agro-industry and the need for appropriate government policies, etc. Only a carefully planned sustainable agriculture can provide a sustainable livelihood security for the poor, which should be the foundation for all development programmes. Gandhiji called such an approach to the development as 'Antyodaya' model, which even foreigners64 are talking about and in its operational sense implies that the priorities in development should be measured by their potential benefit to the poorest sections of the community. In this exercise food of course will come as the first priority and in countries like India it can be produced only by the judicious use of fertilizers.

As regards environmental problems we have to find out ways to overcome these and reduce pollution hazards. A number of techniques are used by agricultural scientists for reducing nitrate leaching losses. These include growing high-yielding crop varieties, deep placement and split application of nitrogen fertilizers, use of ammonium or amide fertilizers and adopting an integrated plant nutrient system (IPNS).

Dudal and Roy defined IPNS as an approach which adapts plant nutrition to a specific farming system and particular yield targets, the physical resource base, the available plant nutrient sources and the socio-economic background. The sources of plant nutrients may be mineral fertilizers and/or biological nitrogen fixation and/or organic materials depending upon a particular location. The FAO-IFFCO International Seminar on IPNS recommended that IPNS should be science-based, associating agronomy, ecology and social sciences. It should use a farming system approach and not limit itself to cropping systems only. It should address both increased productivity and profitability and integrate maintenance and rehabilitation of natural resources.

Some new and promising methods include the use of nitrification inhibitors and slow-release nitrogen fertilizers including indigenous materials such as neem cake or oil-coated materials. Some of these newly developed ecofriendly fertilizers are expensive and their use in field crop production may not be economic. The question is 'should the farmers pay for environmental protection'. After all they are producing food for others, without which this country will have to import and incur heavy expenses. To conclude I would like to quote two sentences from Nobel Laureate Borlaug's keynote address at the 15th World Congress of Soil Science: `Indeed for those concerned with trying to preserve pristine environments or protect endangered species, we would submit that human demographic changes are the greatest threat to the planet Earth in the years ahead. Indeed, if this relentless growth in human numbers goes on unabated, Homo sapienswill no doubt end up as an endangered species themselves'. Fertilizers cannot check this growth in population but can help the world and its nations meet their increasing food, feed and fibre demands and in sustaining the humanity. Thus in developing countries reeling under population pressure the sustainable agriculture and efficient fertilizer use must go hand-in-hand for a better tomorrow.

Technologies: Sustainable Farming

India - Policies for Sustainable Agriculture

The Indian government's policies have always emphasized food grain self-sufficiency, which has not necessarily coincided with agricultural sustainability. The growth of agricultural production and productivity, which had risen significantly during 1970s and 1980s, declined during 1990s. These slowdowns have worsened since 2000; both overall agricultural production and food grains production have shown negative growth rates in 2000-01 to 2002-03 periods (GoI, 2002). Decline in the growth rates of agricultural production and productivity is a serious issue considering the questions of food security, livelihood, and environment. As such, a critical examination of the approaches for sustainable agricultural development is necessary. This examination must be framed not only by India's ongoing need to ensure food self-sufficiency but also by the consequences of access to international markets.

Environmental Challenges in Indian Agriculture

The challenge for Indian agriculture, to put simply, is to increase production, while minimizing environmental impact. This includes conserving and protecting the quality of the resources that determine the performance of agriculture like land, water and air. Reductions in yield, although determined by many factors, may be partially a consequence of land and water exploitation.

Land degradation is one major constraint for Indian agriculture. By the early 1980s approximately 53 percent (173.6 million hectares) of India's geographical area had been considered degraded according to the Ministry of Agriculture (GoI, 2001a): Water logging affected about 6 percent of the cultivated area, while alkali and acidic soils both affected about 3 percent. The major process of land degradation is soil erosion (due to water and wind erosion) contributing to over 71 percent of the land degradation (GoI 2001a). Data compiled by the National Remote Sensing Agency (NRSA) indicated that 15 percent of India's total geographical area was comprised of degraded cultivatable wasteland (NRSA 2000). One third of this land was degraded by human activities, while nearly one half was degraded by a combination of human and natural causes (NRSA 2000). Chadha et al. (2004) found a negative and significant negative relationship between land degradation and foodgrain productivity in both the 1980s and 1990s.

Water is another major constraint for Indian agriculture. Agriculture, through irrigation, accounted for 83 percent of the total water use in the country during 1990 (Vyas 2003). During the Green Revolution period water consumption in agriculture rose sharply as the net irrigated area increased from 31.1 to 54.68

million hectares between 1970-71 and 2000-01, while the area irrigated more than once per year increased from 7.09 million to 20.46 million hectares during the same period. Groundwater, one of the India's major sources for irrigation, is being rapidly depleted. The number of dark blocks (taluk or mandals), where groundwater extraction is more than 85 percent of the availability, increased from 253 to 428 out of over 5700 blocks between 1984-85 and 1998-99 (GoI 2002). The problem of groundwater depletion has been reported from rainfed states like Andhra Pradesh, Karnataka, Rajastan, Madhya Pradesh, Chattisgarh and Gujarat.

The introduction of modern technology based agricultural systems, in addition to encouraging increased water usage, meant the application of inputs like chemical fertilizers, chemical pesticides and high yielding varieties. Fertilizer application rose more than five-fold between 1970 and 2002 to 17360 thousand tonnes. Imbalanced proportioning of chemical nutrients is a major problem associated with fertilizer application in India. Pesticide consumption increased from 24.32 million tonnes in 1970-71 to 46.2 million tonnes in 1999-00, with a peak application of 75.42 million tonnes during 1988-89 (CSE 1999). High yielding seed varieties have lead to mono-cropping of certain grains reducing farmers' cropping flexibility and reducing agricultural biodiversity.

Although, the Indian government has recognized the necessity of managing and conserving resources for agricultural development since the First Five Year Plan, the measures initiated have been inadequate. For example, the government's efforts have only been able to regenerate 17.28 percent of the total degraded area (173.6 million hectares; GoI 2001a).

India's National Agricultural Policy (NAP) (GoI 2000) has stressed the importance of management and conservation of resources by stating that, 'the policy will seek to promote technically sound, economically viable, environmentally non-degrading, and socially acceptable use of country's natural resources - land, water and genetic endowment to promote sustainable development of agriculture'. The Central and state governments have initiated several measures to promote sustainable agricultural development. The NAP stated that improving the quality of land and soil, rational utilisation and conservation of water, and sensitizing the farming community to environmental concerns would receive high priority (GoI 2000).

The Tenth Five Year Plan (GoI 2002), for 2002 through 2007, has put emphasis on natural resource management through rainwater harvesting, groundwater recharging measures and controlling groundwater exploitation, watershed development, treatment of waterlogged areas. With regard to application of agricultural inputs like fertilizer and pesticides, the Plan stated that factors such as imbalanced use of nitrogenous (N), phosphatic (P) and postassic (K), increased deficiency of micronutrients and decreased soil organic carbon would be addressed through a holistic agri-environmental approach stressing Integrated Plant Nutrient and Pest Management. Further, the Tenth Plan document recognizes organic farming as a 'thrust area' in the sustainable use and management of resources in agriculture.

Way Forward

The trajectory of Indian agriculture and its associated environmental problems has brought about recognition that future agricultural growth and productivity will have to occur simultaneously with environmental sustainability. The environmental challenges, especially in terms of land degradation and groundwater depletion, water logging and excessive use of chemical inputs are posing problems for the future of Indian agriculture. To address the problems, policies have laid emphasis on promoting sustainable agriculture including organic farming. Differential approaches and policy instruments, however, will be required to address these problems. The shift from input-intensive to sustainable, particularly organic farming is a difficult task as it involves a number of policy measures dealing with a variety of issues ranging from the transfer of information and technology to the development of markets. Another difficult task, and perhaps more difficult, relates to marginal and small farmers - which comprise a substantial part of Indian agriculture. Although these marginal and small farmers have been considered organic by `default', severe resource constraints make a shift to the modern sense of organic farming prohibitive.

Promoting Sustainable agriculture in India

Agriculture sector, world over, has experienced a phenomenal growth since the mid-twentieth century. The growth, driven by Green Revolution technology, has made a significant dent on aggregate supply of food grains, ensuring food security to the growing population. The next stage of agricultural growth however, faces a serious challenge in terms of sustainability. Whereas the main problem faced by the developing countries in the south pertains to sustainability of resource use, the main challenge facing the developed economies in the north is overuse of chemical inputs. These problems have led to increasing awareness and a felt need for moving away from the input intensive agriculture perused during the Green revolution phase, to sustainable farming in different parts of the world. While the need for a paradigmatic shift in the growth strategy is well recognized, the transition from input intensive to sustainable farming however, has certain inherent difficulties. Notwithstanding these limitations, policies in both the north as well as the south have led increased emphasis on promoting sustainable agriculture.

India can safely be characterized as an agricultural country despite the recent spurt in manufacturing and services and the declining share of agriculture in the national income, since majority of its workforce (~ 65%) are still engaged in agriculture and allied activities. It has been the noblest profession in India since the time immemorial and has been carried out on sustainable basis. It is only relatively recent phenomenon that large-scale forest areas, grazing lands and waste lands have been converted into croplands to support the rising population, which has caused ecological imbalance and atmospheric pollution. With no further scope for expansion of agricultural land efforts have been made to enhance the production of food grains using high-yielding variety of seeds, fertilizers and irrigation along with advanced farm equipments. However, so-called green revolution is cofined to a few crops, viz, wheat, rice and maize and has been possible only in restricted areas, i.e., Punjab, Haryana and Western Uttar Pradesh and certain selected districts of Andhra Pradesh, Maharashtra and Tamilnadu.

Naturally much work is needed to lift the agriculture to a level where it is least affected by vagaries of monsoon and needs little from outside the farm, i.e., lesser dependence on chemical fertilizers and water. The limited success of green revolution has been a mixed bag in that it has given rise to new set of problems: overuse of water and fertilizers. Excessive use of water results in water logging and salinization whereas excess of fertilizers and pesticide cause pollution of water bodies contamination of ground water. India has the largest area of irrigated land (55 million hectares) of which about one-third land is already degraded and 7 million hectare have been abandoned1. In such a situation a renewable and lasting alternative, sustainable agriculture, has to emerge for successful agricultural revolution. In the present paper the need for sustainable agriculture has been emphasized. Policies for sustainable agriculture and organic farming and possible actions in India are discussed.

Need for sustainable agriculture

We can compare three broad types of farming: traditional production systems, conventional modern agriculture (such as Green Revolution technologies), and sustainable agriculture. We can compare them across three dimensions: ecological, economic and social.

Ecological sustainability

Many traditional and most conventional farm practices are not ecologically sustainable: they overuse natural resources, reducing soil fertility, causing soil erosion, and contributing to global climatic change. Sustainable agriculture has several major advantages over both traditional and conventional practices:

Soil fertility: A continuous fall in soil fertility is a major problem in many parts of India. Sustainable agriculture improves fertility and soil structure and prevents erosion, so would be an answer to this problem.

Water: Irrigation is the biggest consumer of fresh water, and fertilizer and pesticides contaminate both surface- and groundwater. Sustainable agriculture increases the organic matter content of the topsoil, so raising its ability to retain and store water that falls as rain.

Biodiversity: Sustainable agricultural practices frequently involve mixed cropping, so increasing the diversity of crops produced and raising the diversity of insects and other animals and plants in and around fields.

Pollution: Pesticides are hazardous to human health as well as to the local ecology. Incorrect handling, storage and use of pesticides lead to health and pollution problems. Sustainable agriculture reduces or eliminates the use of hazardous chemicals; instead it controls pests with a variety of biological and agronomic measures and the use of natural substances.

Landscape: Agriculture and forestry clothe the rural landscape. Inappropriate use causes erosion, landslides and flooding, clogs irrigation channels, and reduces the ability of the land to support the local population. Impoverished rural people flock into the cities in search of jobs, forming unsightly, insanitary slums that further destroy the landscape. Rehabilitating ecologically damaged areas needs huge investments that few countries can afford. Sustainable agriculture avoids these problems by improving productivity, conserving the soil, avoiding the expansion of farming into unsuitable areas, and preserving rural jobs.

Climate: The way agriculture is practiced contributes significantly to global climatic changes. Conventional agriculture contributes to the production of greenhouse gases in various ways: by reducing the amount of carbon stored in the soil and in vegetation, through the production of methane in irrigated fields, and through energy-intensive activities such as the production of artificial fertilizers. Adopting sustainable agriculture would reduce these impacts significantly.

Economic sustainability

Agriculture cannot be sustainable unless it is economically viable over the long term. Conventional agriculture poses greater long-term economic risks than "sustainable" alternatives.

Export vs. local orientation: Governments tend to view export-oriented production systems as more important than those that supply domestic demands. This is misguided. Focusing on exports alone involves hidden costs: in transport, in assuring local food security, etc. Policies should treat domestic demand and in particular food security (either by farmers producing food for themselves, or by selling produce for cash they can use to buy food) as equally important to the visible trade balance.

Debt: The Green Revolution raised India's grain output significantly, but a vast number of small-scale farmers ran into a debt trap: they took out loans to raise their production, and then found they could not pay the money back. About 40,000 were so desperate that they committed suicide.

Risk: Concentrating on specific commodities seems to promise high economic returns. But market production implies certain risks: markets change quickly, and international agricultural prices are dropping. Cheap foreign food may sweep into the national market, leaving Indian farmers without a market. As a World Trade Organization signatory, the Indian government is under pressure to deregulate and open its economy to the world market so cannot protect its farmers behind tariff walls.

Niche markets: Organic agriculture is one of the strongest ways to farm in an environmentally sustainable way. The demand for certified organic products is increasing quickly, opening opportunities to expand sales of such products and to explore niche markets.

Employment: Farming is the main source of employment for rural people. Trends towards specialization and mechanization may increase narrowly measured "efficiency", but they reduce employment on the land. The welfare costs of unemployment must be taken into account when designing national agricultural support programmes. Sustainable agriculture, with its emphasis on small-scale, labour-intensive activities, helps overcome these problems.

Social sustainability

The social sustainability of farming techniques is related to the ideas of social acceptability and justice.

Inclusiveness: Development cannot be sustainable unless it reduces poverty for the broad masses of people in India. The government must find ways to enable the rural poor to benefit from agricultural development.

Political unrest: Gaps between the "haves" and "have-nots" feed a feeling of social injustice among those who feel neglected and excluded from development opportunities, as well as from better-off sympathizers. The result is a climate favorable to political opposition and even violence.

Local acceptance: Many new technologies fail because they are based on practices or assumptions from outside. Sustainable agricultural practices usually are based on local social customs, traditions, norms and taboos, so local people are more likely to accept them and adapt them to their own needs.

Indigenous knowledge: Sustainable agricultural practices often rely on traditional knowhow and local innovation. Local people have a wealth of knowledge about their environment, crops and livestock. They keep locally adapted breeds and crop varieties. They have social structures that manage and conserve common resources, help people in need, and maintain the social fabric. Rather than ignoring or replacing this knowledge, sustainable agricultural development seeks to build on it and enrich it with appropriate information from outside.

Gender: In traditional agriculture, women traditionally bear the heaviest burdens in terms of labour. In modern conventional farming, too, men often benefit the most: they control what is grown and how the resulting income is spent. Sustainable agriculture attempts to ensure that the burdens and benefits are shared more equitably between men and women.

Food security: Traditional farming techniques often fail to produce enough food, or enough variety of food for a balanced diet. Conventional modern farming focuses on a few commodities, so people still do not have a balanced diet. Sustainable agriculture improves food security by improving the quality and nutritional value of the food, and by producing a bigger range of produce throughout the year.

Participation: Traditional society in India is raven by wealth and caste distinctions. Introducing conventional farming innovations tends to exacerbate these: the rich and higher-caste tend to benefit, while the poor and lower-caste are left out. Sustainable agricultural interventions consciously target the less well-off, and empower them so they can organize and speak with their own "voice", so promoting dialogue and democracy.

Sustainable Agriculture in India

The sustainable agriculture may be defined as any set of agronomic practices that are economically viable, environmentally safe, and socially acceptable. If a cropping system requires large inputs of fertilizer that leak from the system to pollute ground water, drinking supplies and distant coastal fisheries, the system may be sustainable economically as the long-term supply of fertilizer is stable and the economic cost of fertilizer is easily borne by larger grain production but it is not sustainable environmentally or socially, since it does not cover the cost of environmental damage or social costs. The organic agriculture focuses on "living soil", on optimizing the use of biological processes and on avoiding the use of synthetic chemicals and fertilizers.

Advocates of sustainable agriculture agree with biological focus and hope to reduce but not necessarily eliminate chemical use. In the context of sustainable agriculture another term "alternative agriculture" has been prominently used. Definition of alternative agriculture sheds much light on operational aspects of sustainable agriculture. Any food or fiber production that has a more thorough incorporation of natural processes, reduced use of off-farm inputs with less harm to environment and consumers, a more productive use of biological and genetic potential of plants and animals, a better match between cropping patterns and the physical capacity of lands and,

An improved emphasis on conservation of soil, water, energy and biological resources, is defined as alternative agriculture.

The normal agricultural practices using irrigation, chemical fertilizer, pesticides and high-yielding variety of seeds is called conventional agriculture. With increasing use of chemical fertilizers and pesticides the conventional agriculture is major source of pollution of inland water bodies and coastal seas. There has been growing criticism of conventional agriculture for its side effects, the "external costs" which impact communities, the environment, and human health.

As for indicators of sustainability there is no single prescription. Sustainable practices will vary by cropping system, local environment and socio-economic system. Still, experience tells us that locally sustainable systems tend to be more resource conservative than less sustainable system and tend to rely less on external inputs and more on internal ecosystem services.

India - Policies for Sustainable Agriculture and Organic Farming

The Indian government's policies have always emphasized food grain self-sufficiency, which has not necessarily coincided with agricultural sustainability. The growth of agricultural production and productivity, which had risen significantly during 1970s and 1980s, declined during 1990s. These slowdowns have worsened since 2000, both overall agricultural production and food grains production have shown negative growth rates in 2000-01 to 2002-03 period. Decline in the growth rates of agricultural production and productivity is a serious issue considering the questions of food security, livelihood, and environment. As such, a critical examination of the approaches for sustainable agricultural development is necessary. This examination must be framed not only by India's ongoing need to ensure food self-sufficiency but also by the consequences of access to international markets.

Sustainable Agriculture - Possible Actions in India

1. Improvement of existing production systems (e.g. altered crop rotations, introduction of green manuring, use of plant species adapted to specific locations)
2. Improved protection of natural resources (e.g. erosion protection)
3. Increase in efficiency of existing resources (e.g. irrigation, use of technology, basic and advanced training)
4. Introduction of regenerative branches of business (e.g. horticulture or aquaculture)

5. Introduction of a new production element in existing enterprises (such as fruit trees to stabilize terraced fields, fish-farming in rice fields)
6. Optimization of post-harvest systems (e.g. storage)
7. Increase the value of agricultural products through further processing (e.g. production of yoghurt from milk)
8. Improvement of channels of distribution (e.g. market access, transport)
9. Access to loans and other financial services
10. Covering risk (e.g. through land law, support of producer groups)

BIBLIOGRAPHY

- Agricultural engineering Britannica Online. Retrieved 2012-12-25.
- Tractor (vehicle) Britannica Online. Retrieved 2012-12-25.
- "How to Manage Prevent Plant Weeds". Successful Farming. 2019-12-06. Retrieved 2020-08-13.
- "New High-Tech Farm Equipment Is a Nightmare for Farmers". Wired. 2015-02-05.
- "We Can't Let John Deere Destroy the Very Idea of Ownership". Wired. 2015-04-21.
- Exemption to Prohibition on Circumvention of Copyright Protection Systems for Access Control Technologies http://copyright.gov/1201/2015/fedreg-publicinspectionFR.pdf
- Farmlabs.org. "This is the Beginning of a network of open laboratories for agricultural research and experimentation". farmlabs.org. Retrieved 2019-07-20.
- Gaillard, Chris. "L'Atelier Paysan". L'Atelier Paysan (in French). Retrieved 2019-07-20.
- Chance, Quentin; Meyer, Morgan (2017-06-06). "L'agriculture libre. Les outils agricoles à l'épreuve de l'open source". Techniques & Culture. Revue semestrielle d'anthropologie des techniques (in French) (67): 236–239. doi:10.4000/tc.8534. ISSN 0248-6016.
- "Ekylibre". ekylibre.com. Retrieved 2019-07-20.
- "Group Overview ‹ Open Agriculture (OpenAg)". MIT Media Lab. Retrieved 2019-07-20.

- "Project Overview ‹ Open Phenome Project". MIT Media Lab. Retrieved 2019-07-20.
- "recipe:start [OpenAg]". wiki.openag.media.mit.edu. Retrieved 2019-07-20.
- 1992. p. 7-8
- Jan H. Maas (1994) De Nederlandse agrosector: geografie en dynamiek p.177
- Gifford (1992, p. 9) even stated that "nearly all farm machinery manufacturers today can be trace their beginnings to a blacksmith."
- Ganzel, Bill (2003). "Tractors". Farming in the 1930s. Wessels Living History Farm.
- Collins, Edward JT. "The rationality of'surplus' agricultural labour: mechanization in English agriculture in the nineteenth century." The Agricultural History Review (1987): 36-46.
- Dewey, Peter. "The British Agricultural Machinery Industry, 1914-1939: Boom, Crisis, and Response." Agricultural History (1995): 298-313.
- McNeil, Ian (1990). An Encyclopedia of the History of Technology. London: Routledge. ISBN 0-415-14792-1.
- Hounshell, David A. (1984), From the American System to Mass Production, 1800–1932: The Development of Manufacturing Technology in the United States, Baltimore, Maryland: Johns Hopkins University Press, ISBN 978-0-8018-2975-8, LCCN 83016269, OCLC 1104810110
- Wells, David A. (1891). Recent Economic Changes and Their Effect on Production and Distribution of Wealth and Well-Being of Society. New York: D. Appleton and Co. ISBN 0-543-72474-3. RECENT ECONOMIC CHANGES AND THEIR EFFECT ON DISTRIBUTION OF WEALTH AND WELL BEING OF SOCIETY WELLS.
- Ford, Henry; Samuel (1922). My Life and Work: An autobiography of Henry Ford.
- Constable, George; Somerville, Bob (2003). A Century of Innovation: Twenty Engineering Achievements That Transformed Our Lives, Chapter 7, Agricultural Mechanization. Washington, DC: Joseph Henry Press. ISBN 0-309-08908-5.[page needed]
- White, William J. "Economic History of Tractors in the United States". Archived from the original on 2013-10-24.
- Ayres, R. U.; Ayres, L. W.; Warr, B. (2002). "Exergy, Power and Work in the U. S. Economy 1900-1998, Insead's Center For the Management of

Environmental Resources, 2002/52/EPS/CMER" (PDF). Archived from the original (PDF) on 2017-08-13. Retrieved 2010-10-06.

- Moore, Stephen; Simon, Julian (Dec 15, 1999). "The Greatest Century That Ever Was: 25 Miraculous Trends of the last 100 Years" (PDF). The Cato Institute: Policy Analysis, No. 364. Fig 13.
- Wall Street Journal: "Robots Step Into New Planting, Harvesting Roles - Labor shortage spurs farmers to use robots for handling delicate tasks in the fresh-produce industry" By ILAN BRAT April 23, 2015
- Washington State University Extension - School of Economic Sciences: "Asparagus Production Economics" December 8, 2010
- Vegetable Growers News: "Mechanical Asparagus Harvester Almost a Reality" December 21, 2009
- Fox Business News: "Machinery takes the place of migrants as Maine's blueberry harvest booms" September 06, 2015
- Albuquerque Journal: "Chile harvester gets a field test in NM" By Diana Alba Soular September 22, 2014
- Current Argus: "Editorial: Trial could be a boon to chile farmers" Archived 2015-08-11 at the Wayback Machine July 31, 2015
- KJZZ: "New Mexico Researchers To Test Mechanized Green Chile Harvesting" by Carrie Jung July 29, 2015
- Las Cruces Sun: "Experts: Machines could reverse declining New Mexico green chile acreage" By Diana Alba Soular Archived 2015-08-10 at the Wayback Machine July 25, 2015
- University of California Davis Migration Files: "The Status of Labor-saving Mechanization in Fruits and Vegetables" By Wallace E. Huffman May 25, 2010
- Farm 2 ranch magazine: "Labor concerns lead peach growers to look to machines" August 18, 2016
- U. S. Produce Industry and Labor: Facing the Future in a Global Industry By Linda Calvin retrieved September 28, 2013
- Fresno Bee: "New raisin grape holds promise for central San Joaquin Valley growers" By Robert Rodriguez September 19, 2015
- Japan Times: "Latest robot can pick strawberry fields forever" September 26, 2013
- "No Hands Touch the Land: Automating California Farms" (PDF). California Agrarian Action Project: 20–28. July 1977. Retrieved 2015-04-25.

- Keith Cowling, David Metcalf, A. J. Rayner. (1973) Resource Structure of Agriculture: An Economic Analysis, p. 127-8
- World bank (1954) Western Europe - Tractor and agricultural machinery industry.
- VDMA. Agricultural Machinery November 2013 : Market Perspectives 2014[permanent dead link], 2013-14. p. 5
- Ade Asefeso (2012). CEO Guide to Doing Business in Australia, p. 54
- Kenneth James, Narongchai Akrasanee. Small and Medium Business Improvement in the ASEAN Region: Production Management. Institute of Southeast Asian Studies, 1 dec. 1988
- The leading machinery manufacturers worldwide in 2012, based on revenue. Accessed 25.11.2014.
- Number of employees of John Deere (Deere & Company) from 2002 to 2013. Accessed 25.11.2014.
- Revenue of tractor & agricultural machinery manufacturing (NAICS 33311) in the United States from 2009 to 2014 (in billion U.S. dollars). Accessed 25.11.2014.
- Yuan'en, Gao (2007). "Agricultural Machinery Industry in China" (PDF). Chinese Association of Manufacturers of Agricultural Machinery.
- Global and Chinese Agricultural Machinery Industry Report, 2013-2014, December 2013. Summary (online). Accessed 25.11.2014.
- Singh, Gyanendra (2006). "Agricultural Machinery Industry in India (Manufacturing, marketing and mechanization promotion)" (PDF). Status of Farm Mechanization in India. Archived from the original (PDF) on March 18, 2013.
- Singh, Sukhpal (April 2009). "Agricultural Machinery Industry in India: A Study of Growth, Market Structure, and Business Strategies" (PDF). Indian Institute of Management Ahmedabad. Archived from the original (PDF) on 2015-09-24.
- "Agricultural Machinery Industry in Turkey". blog.tcp.gov.tr. Turkish Ministry of Economy. 1 June 2011. Archived from the original on December 5, 2014.

INDEX

9 789360 843359

Printed at Repro India Ltd.